Liquid Crystals

Liquid Crystals

NATURE'S DELICATE PHASE OF MATTER

Second Edition

Peter J. Collings

PRINCETON UNIVERSITY PRESS

PRINCETON AND OXFORD

Copyright © 1990, 2002 by Princeton University Press
Published by Princeton University Press, 41 William Street, Princeton, New Jersey 08540
In the United Kingdom: Princeton University Press,
3 Market Place, Woodstock, Oxfordshire OX20 1SY
All Rights Reserved. First edition 1990. Second edition 2002.

Library of Congress Cataloging-in-Publication Data

Collings, Peter J., 1947–
    Liquid crystals : nature's delicate phase of matter,
    2nd ed. / Peter J. Collings.
  p. cm.
Includes bibliographical references and index.
ISBN 0-691-08672-9 (alk. paper)
  Liquid crystals.

QD923 .C638 2002   2001050027

British Library Cataloging-in-Publication Data is available

This book has been composed in Times

Printed on acid-free paper. ∞

www.pup.princeton.edu

Printed in the United States of America

1 3 5 7 9 10 8 6 4 2

In memory of my mother, Juanita B. Collings

———————————

CONTENTS

PLATES

PREFACE TO THE SECOND EDITION

IT HAS BEEN twelve years since the first edition was written. During this period of time, countless people have told me how useful the book was as they started to perform research on liquid crystals or as they satisfied their curiosity about this fascinating state of matter. But science never stands still, and as a result we have gained new knowledge about liquid crystals during this time. In most cases this has meant a recognition that this phase of matter is even richer in phenomena than we had thought. Understanding this newly discovered behavior has usually refined our general ideas concerning liquid crystals and in some cases has even forced us to replace previous explanations with new ones. In order to provide both a balanced and up-to-date introduction into the liquid crystalline phase, therefore, the first edition needed to be revised and enhanced.

Most of the changes in the second edition take the form of additional information that has been acquired since the writing of the first edition. These new sections have been added at appropriate places, leaving the overall structure of the book unchanged. Some of these new subjects are (1) an up-to-date general discussion of liquid crystal research, (2) coverage of the anti-ferroelectric chiral smectic phase and its application in displays, (3) a more extensive section on freestanding liquid crystal films and the layering transitions that take place in them, (4) additional examples of displays incorporating both polymers and liquid crystals, (5) a much deeper discussion of structure-property relations in thermotropic liquid crystals, (6) a section devoted to the more highly ordered smectic phases and the recently discovered bent-core or "banana-shaped" liquid crystals, (7) new sections on nematic emulsions and the liquid crystal phases that form in self-assembling (supramolecular) systems, (8) additional description of structure-property relations in polymer liquid crystals, including elastomers, (9) an additional section on twist-grain-boundary phases, (10) new information made possible from recent computational work, and (11) an updated section on biological work involving liquid crystals. In addition to these content changes, new figures have been added, and all the figures have been redrawn using computer software. It is hoped that these changes will serve to make the second edition an even more useful introduction to the chemistry and physics of the liquid crystal phase than the first edition was.

The reason liquid crystal research is an important and active field of science has changed over the last twelve years. In the 1980s, just understanding the wide range of phenomena associated with liquid crystals involved new science. New substances showed new properties, and the new theoretical ideas necessary to understand these properties had implications beyond the area of liquid crystals. In addition, the market for lightweight and low-power computer displays was growing exponentially, fueling a boom in the application of liquid crystals to displays. In the 1990s, the rate at which discoveries concerning liquid crystals revealed new scientific ideas slowed considerably, but at the same time it was realized that liquid crystals had roles to play in systems very different from the simple ones containing only liquid crystal compounds. Liquid crystals were mixed with polymers and nonliquid crystal-forming liquids, incorporated into thin films, probed as the result of self-assembling systems, and used to create unique materials in many different ways. Talks at scientific conferences started to take on a different perspective, not as much concerned with new findings in simple liquid crystal systems, but more describing the interesting and sometimes unexpected behavior of more complex systems. Liquid crystal research is still thriving, but in line with its interdisciplinary roots, it is branching out to include people with a far wider set of specialties than ever before. The "golden years" of the 1980s might be over, but the richness and excitement in fields where liquid crystals play an important part is as high as ever.

PREFACE TO THE FIRST EDITION

I HAVE HEARD it argued, usually by nonscientists, that to study nature using the methods of science tends to diminish the glory of the natural world in which we live. The claim is that nature is somehow "demystified" and simplified by science. I doubt very much if a scientist would ever argue in this way. Any person who has seriously examined what we know about the inner workings of a cell or carefully considered the theory of matter during the first minutes of our universe can only stand in awe of nature's elaborate yet functional design. Even old discoveries never seem to lose their charm. Each new hypothesis or experimental result has a way of affecting everything that came before it, sometimes in a dramatic way and often only in a subtle way. Our view of nature is constantly in flux, with new revelations occurring in both expected and unexpected areas. Scientists rest only momentarily to celebrate knowing something new about nature, in that this new "worldview" always changes what they thought they knew and forces them to ask questions they never would have thought to ask. This scientific perspective on the natural world is both beautiful and glorious, but not in a mysterious way. It is more analogous to the relationship between two good friends. Rather than being a detriment to the relationship, the increasing familiarity in a friendship allows both people to develop in ways not possible individually.

Such is the case with science and nature. Nothing exemplifies these ideas better for me than my studies involving liquid crystals. I have been engaged in liquid crystal research for over sixteen years, and as a result know much more about them than when I started. Nevertheless, they never seem to stop amazing me. I recall very vividly the first time I actually saw a substance form a liquid crystal phase. I had been learning about liquid crystals for a few months, and understood many of their properties. Yet observing what I had only previously imagined added meaning to what I had learned and made me want to learn more. The truth is that I continue to have those same feelings at times in the laboratory. It still happens once in a while as I watch a substance form a liquid crystal phase under the microscope, but more often I get that feeling when an experiment reveals that the liquid crystal does not behave as our theories predict. All I can think about is why, and I want to learn more.

In writing this book, I could not avoid attempting to convey some of these feelings of wonder and beauty. I have also tried to include as much background information as necessary, so that a reader need not consult other references in order to understand the concepts. Hopefully this will allow the reader to learn more and more about liquid crystals, while at the same time finding enjoyment in understanding this beautiful phase of matter. I have also tried to be as comprehensive as possible, covering all the important areas in biology, chemistry, and physics. In doing this, I have tried to keep the scientific level consistent throughout the book. A reader familiar with the basic ideas of biology, chemistry, and physics should have little difficulty in following the explanations and arguments. There is a short chapter on the discovery and early history of liquid crystals, as well as a chapter that discusses some of the more modern and theoretical research being performed at this time. Two chapters discuss many of the important technical applications of liquid crystals, with appropriate emphasis on the liquid crystal display. Finally, ample space is devoted to the exciting work being done with polymer liquid crystals, liquid crystal emulsions, and liquid crystals in biological systems.

ACKNOWLEDGMENTS

MANY PEOPLE have provided important assistance during the writing of the second edition of this book. In particular, I would like to thank Scott Collings for producing most of the figures in this edition, all my colleagues in the Department of Physics and Astronomy at Swarthmore College for providing the rich intellectual climate that makes teaching and research such natural endeavors, and all the students I have taught or supervised, for without them my excitement for science would not have endured. Finally, I thank the following people for providing pictures for the book: W. Becker (E. Merck Company), W. Blake (R. H. Blake, Inc.), P. Crooker, J. Diamond (Eastwick Communications), B. Dribben (Flexitherm Thermographic Systems), J. Goodby, H. Kitzerow, A. Lowell (Samsung Electronics), M. Neubert, E. Samulski, H. Stegemeyer, M. Wand (Displaytech), and D. Weiss.

Liquid Crystals

Chapter 1

WHAT ARE LIQUID CRYSTALS?

ONE FAMOUS SCIENTIST once remarked that he found liquid crystals to be quite mysterious. Certainly the name itself is confusing. After all, how can something be both liquid and crystalline? We shall see in this chapter that in some cases a very interesting compromise spontaneously occurs in the natural world, and that the name used to describe the result of this compromise is quite appropriate. It will also be evident that liquid crystals come in a number of forms, creating an area of study rich in new and exciting phenomena.

STATES OF MATTER

We are all aware of the fact that many substances can exist in more than one state of matter. The most familiar example is water, which is a solid below 0°C (32°F), a liquid between 0 and 100°C (212°F), and a gas above 100°C. Solids, liquids, and gases are the most common states of matter, but in spite of what many children learn in school, they are not the only states of matter.

These three common states of matter are different from each other because the molecules in each state possess different amounts of order. The solid state consists of a more or less rigid arrangement of molecules because each molecule occupies a certain place in the arrangement and remains there. Not only are the molecules constrained to occupy a specific position, but they are also oriented in a specific way. The molecules might vibrate a bit, but on average they constantly maintain this highly ordered arrangement. There are large attractive forces holding the molecules of a solid in place, because the arrangement causes the forces between individual molecules to add together. It therefore takes large external forces to disrupt the structure, so solids are hard and difficult to deform.

The liquid state is quite different in that the molecules neither occupy a specific average position nor remain oriented in a particular way. The molecules are free to diffuse about in a random fashion, constantly bumping into one another and abruptly changing their direction of motion. The amount of order in a liquid is therefore much less than in a solid. Attractive forces still exist in a liquid, but the random motion of the molecules does not allow the forces between individual molecules to add together. The result is that the forces holding liquid molecules together tend to be much weaker than the forces in solids. These forces are strong enough, however, to keep the molecules fairly close to each other. A liquid, therefore, maintains a constant density, even though it takes the shape of its container. The lack of a rigid arrangement allows liquids to be deformed quite easily. Liquids flow and will change their shape in response to weak outside forces. The fact that the molecules are held close together is evident when one tries to compress a liquid. Liquids are difficult to compress, a characteristic that is put to successful use in the hydraulic systems of automobile brakes and earth-moving equipment.

In the gaseous state, the more chaotic motion of the molecules causes the attractive forces between them to add together even less than in the liquid state. The amount of order is therefore less than in liquids, and the forces are not strong enough to hold the molecules close together. The molecules move about in the same fashion as in liquids, but they eventually spread evenly throughout the container no matter how large it is. Thus, the liquid and gaseous states are very similar; the motion of the molecules is random and disorganized in both states. But in the liquid state the molecules maintain a specific average distance between each other, while in the gaseous state the average intermolecular distance is determined by the number of molecules and the size of the container. A gas can be deformed even more easily than a liquid. In fact, a gas can be significantly compressed since it takes much less force to move the molecules a little closer together.

I should mention that the description I have given for the solid state is really appropriate for a crystalline solid. Some substances in the liquid state change into an amorphous solid rather than a crystalline solid when cooled. The molecules in an amorphous solid are fixed in place, but there is no overall pattern to the arrangement of these molecules. The molecules are arranged more or less randomly (similar to a snapshot taken in a liquid), but unlike in the liquid state, they do not diffuse throughout the substance. Some of the substances discussed in this book possess crystalline solid states while others possess amorphous solid states.

To understand why different substances form certain phases, we must consider the effects due to temperature. Temperature is a measure of the random motion of molecules. The higher the temperature, the more the molecules are moving and vibrating in a random way. Since the attractive forces between the molecules of a substance in a certain state of matter do not change with temperature (although the random motion of molecules does increase with temperature), the ability of the attractive intermolecular forces to keep the molecules ordered in any way must decrease as the temperature increases. Consider water for example. Below 0°C the attractive forces between the water molecules when arranged so rigidly are strong enough to hold the molecules firmly in place, even though they all possess random motion due to the temperature. Above 0°C, however, the random motion becomes too violent and the intermolecular forces cannot hold the molecules in place, causing the solid to melt. The random motion causes the molecules to wander around, which reduces the forces holding the molecules together because the forces between molecules no longer add together as in the solid state. These forces, however, are still great enough to keep the molecules from wandering completely away from each other. The liquid takes up a specific amount of space. At a temperature above 100°C the random motion is so violent that the attractive forces are not even capable of keeping the molecules next to one another. The water is now in the gas phase, and the intermolecular forces are even weaker than in the liquid phase, because the forces between molecules add together even less. The molecules spread out to fill the available volume.

I can now make a generalization from our discussion concerning water. Every substance possesses intermolecular forces of some kind, so at any temperature, a substance exists in a specific state of matter. Scientists also use the word *phase* to describe a specific state of matter, and refer to the specific state as the stable phase at that temperature. For example, water at 20°C (68°F) is stable in the liquid phase. Carbon dioxide is stable in the gas phase at this temperature, and salt (sodium chloride) is stable in the solid phase. The intermolecular forces are different in these three substances, in each case causing a different phase to be stable at the same temperature.

When the temperature is changed so that the phase that was stable no longer is stable, the substance changes phase. The change of phase occurs at a precise temperature because at that temperature the ability of the inter-molecular forces to cause that phase to exist is no longer sufficient. We say a *phase transition* occurs at that temperature. The important change that takes place at a phase transition is the amount of order among the molecules of the substance.

In most cases the process is reversible. A liquid to solid phase transition occurs at 0°C for water if it is being cooled. The fact that any substance has phases that are stable over certain temperature ranges allows us to draw a single diagram for each substance summarizing its phase behavior. A horizontal line represents temperature, with higher temperature to the right. Phase transitions are indicated by short vertical lines at the proper temperature, and the names of the stable phases are written between the appropriate phase transition lines. The diagram for water is shown in figure 1.1.

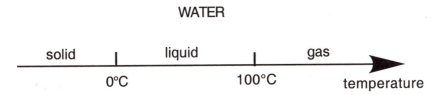

Fig. 1.1. Phase diagram of water. The stable phases of water are given above the temperature axis, with the phase transition temperatures shown below the axis.

Before I proceed, let me digress and mention a fourth state of matter called the *plasma* phase. It has nothing to do with liquid crystals, but it is a true state of matter just as the solid, liquid, and gaseous states are. If a substance is heated to a high enough temperature (and I do mean very high), the random motion becomes so violent that electrons that are normally bound to the atoms get knocked off. This phase of matter is composed of positively charged ions (atoms that have lost one or more electrons) and negatively charged electrons, which normally attract each other so strongly that the ions and electrons bind together. The temperature is so high, however, that the rate the ions and electrons bind together is equal to the rate the electrons are being knocked off atoms. Thus the substance exists in this state with unbound ions and electrons. It is a new phase of matter, one that normally exists in and around stars. Scientists presently create a plasma in their experiments on nuclear fusion. The temperatures of these plasmas are so high that scientists continue to learn new things about how matter behaves under these conditions.

THE LIQUID CRYSTAL PHASE

I will now complicate this picture one more time and discuss liquid crystals. Instead of doing an experiment with water, let's imagine that we have a substance called cholesteryl myristate and wish to investigate its phases. Cholesteryl myristate is a complicated molecule, composed mostly of carbon and hydrogen atoms. It is a fairly common substance, however, and can be found in our cell membranes and also in those nasty deposits that cause hardening of our arteries. At room temperature (20°C) it is in the solid phase, so we heat it up slowly and observe what takes place. At 71°C the solid melts, but the resulting "liquid" is very cloudy and not like other liquids such as water, alcohol, or cooking oil. If we keep raising the temperature, we notice that another change takes place at 85°C; here the cloudy "liquid" turns clear, now looking like other familiar liquids. We continue to heat up the substance; nothing further happens, and we reach the maximum temperature of our apparatus (200°C).

Based on what I have discussed, you may be thinking that these sharp changes at a specific temperature must be phase transitions. If so, the cloudy "liquid" must represent a phase different from both the solid and the liquid phases. This idea is indeed correct, and the phase is called the *liquid crystal* phase. It is a fluid phase in that a liquid crystal flows and will take the shape of its container. Its cloudiness, however, indicates that it differs from liquids in some fundamental way. It was no simple job for scientists of the late nineteenth and early twentieth centuries to put together a proper conception of this new state of matter. But through lots of experimentation and some good creative thinking, a useful picture of the liquid crystal phase began to emerge. Before going any further, let us draw a diagram showing the phases of cholesteryl myristate (see figure 1.2). I am certain a gas phase of cholesteryl myristate exists at some high temperature, but the molecules begin to break apart at high temperatures. Therefore, it might be impossible to obtain a gas phase of the pure compound.

The molecules in a solid are constrained to occupy only certain positions. We describe this condition by saying that the solid phase possesses *positional order*. In addition, the molecules in these specific positions are constrained in the ways they orient themselves with respect to one another. We say the solid phase also possesses *orientational order*. When a solid melts to a liquid, both types of order are lost completely; the molecules move and tumble randomly. When a solid melts to a liquid crystal, however, the positional order may be lost although some of the orientational order remains. The molecules in the liquid crystal phase are free to move about in much the same fashion as in a liquid, but as they do so they tend to remain oriented in a certain direction. This orientational order is not nearly as perfect as in a solid; in fact, the molecules of a liquid crystal spend only a little more time pointing along the direction of orientation than along some other direction. Still this partial alignment does represent a degree of order not present in liquids and thus requires that we call this condition a new phase or state of matter. Figure 1.3 illustrates the order present in the solid, liquid crystal, and liquid phases.

How do we quantitatively describe the amount of orientational order present in a liquid crystal? Since the molecules are not fixed, we are forced to describe the order as some sort of average. For example, let us imagine that the direction of preferred orientation in a liquid crystal is toward the top or bottom of the page. This direction can be represented by a vertical arrow, called the *director* of the liquid crystal. There are always two choices as to which way the director points (up or down in this example). The two directions are equivalent in the liquid crystal, so either choice is fine. One way we could visualize performing an average is to take a snapshot of a rep-

CHOLESTERYL MYRISTATE

Fig. 1.2. Phase diagram of cholesteryl myristate. A gas phase is not shown, because the molecule decomposes at high temperatures.

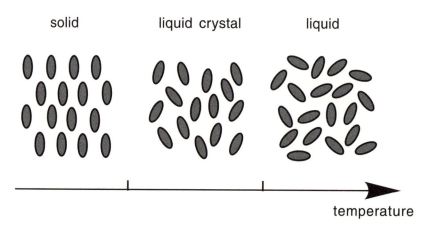

Fig. 1.3. Schematic illustration of the solid, liquid crystal, and liquid phases. The elliptical shapes represent molecules.

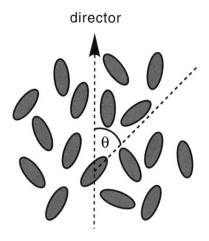

Fig. 1.4. A snapshot of the molecules in the liquid crystal phase. The dashed arrow (director) shows the direction of preferred orientation. Each molecule makes an angle with the director (shown for one molecule as θ).

resentative group of molecules at a certain time. Each molecule in our picture is oriented at some angle relative to the director, so our snapshot might look something like figure 1.4. We could measure all the angles and compute the average angle as a measure of the amount of orientational order. The more orientational order present, the closer the average angle would be to zero. A snapshot of the molecules in a liquid with no orientational order would yield values for the angle between 0° and 90°, with angles representing all possible directions distributed at random. We must keep in mind that this random arrangement occurs in three dimensions. Innumerable orientations make an angle of 90° to the director, but only one makes an angle of 0°. There will therefore be many more molecules making angles of 90° with the director as compared to 0°. If we average these measurements, the larger number of molecules oriented at larger angles produces a result greater than 45° (57° to be exact). Using this scheme, therefore, no orientational order means an average angle of 57°, with smaller angles indicating the presence of orientational order. The most order possible (complete alignment) would give an average angle of 0°.

There is nothing wrong with this procedure, but for many reasons a different method is more useful. In this new procedure, the angle the molecule makes with the director is not averaged; instead, the function $(3\cos^2\theta - 1)/2$ is averaged. Since the cosine of 0° is 1, perfect orientational order (all angles equal to 0°) causes this average to equal 1. In addition, in a liquid with no orientational order, the average of this function is 0. This is a more meaningful range of values, and the average of this function is called the *order parameter* of the liquid crystal. It is an extremely important quantity. The order parameter of a liquid crystal decreases as the temperature is increased and, as can be seen in figure 1.5, typical values for the order parameter are between 0.3 and 0.9.

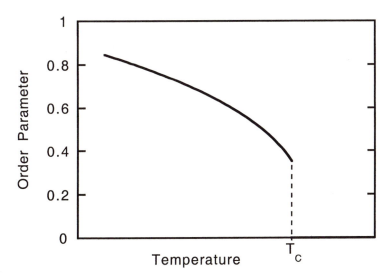

Fig. 1.5. Order parameter variation with temperature in the liquid crystal phase. T_C represents the transition temperature to the liquid phase.

We could have performed this average in another way. Imagine following a single molecule, taking pictures of it at regular intervals. You can imagine measuring the angle in each of these pictures and using them all to perform the average. Do you expect that the averages performed in these two different ways would yield different values for the order parameter? The answer is no if you assume that all molecules undergo the same type of random motion. If this assumption is true, the orientation one molecule has now is the same as another molecule has at some other time. Performing an average of one molecule's orientation at many times should be equivalent to performing an average of many molecules' orientations at any one time. Interestingly enough, in all the experiments performed on liquid crystals over the past one hundred years, none of the results have ever contradicted this assumption.

Is a liquid crystal more like a solid or a liquid? This question is important because it turns out that there is a definite answer. When a phase transition takes place from solid to liquid, for example, energy must be supplied to the substance to disrupt the attractive forces that hold the solid together in a highly ordered arrangement. An analogy is the energy you must supply in order to pull two magnets apart. If ice is the substance, you must supply 80 calories of energy for each gram of ice in order to melt it. Likewise, it takes energy to pull the molecules of a liquid away from each other in changing to the gas phase. For example, a gram of water must be supplied with 540 calories of energy before it will all be changed to steam. The amount of energy required to cause a phase transition is called the *latent heat* of the transition and is a useful measure of how different the two phases are. In the case of cholesteryl myristate, the latent heat of the solid to liquid crystal transition is 65 calories/gram, while the latent heat for the liquid crystal to liquid transition is 7 calories/gram. These numbers allow us to answer the question posed earlier. The smallness of the latent heat of the liquid crystal to liquid phase transition is evidence that liquid crystals are more similar to liquids than they are to solids. When a solid melts to a liquid crystal, it loses most of the order it had and retains only a bit more order than a liquid possesses. This small amount of order is then lost at the liquid crystal to liquid phase transition. The fact that liquid crystals are similar to liquids, with only a small amount of additional order, is the key to understanding the many physical properties that make them nature's delicate state of matter.

Although most molecules do not form a liquid crystal phase, it is not a rare occurrence when one does. It is said that an organic chemist indiscriminately synthesizing compounds would find out that about one in every two hundred possesses the liquid crystal phase. After years of experiments, it has become clear what type of molecule is likely to be liquid crystalline at some temperature. First of all, the molecule must be elongated in shape; that is, it must be significantly longer than it is wide. Second, the molecule must have some rigidity in its central region. A molecule that flops around like a piece of cooked spaghetti is unlikely to have a liquid crystal phase. Finally, it seems to be advantageous if the ends of the molecule are somewhat flexible. A good model of a typical liquid crystal molecule is therefore a short pencil with a short piece of cooked spaghetti attached to each end. Why this model works is not difficult to understand. Elongated molecules usually have stronger attractive forces when they are aligned parallel to one another. In addition, elongated molecules bump into each other less when they all tend to point in the same direction, a fact that acts to stabilize aligned phases. The importance of the flexible ends is a bit more subtle. The flexibility seems to allow one molecule to position itself more easily between other molecules as they all move about.

Although it is possible to observe and identify liquid crystals in bulk samples, one of the most useful and beautiful ways to observe this phase is under a microscope. First, some liquid crystal is placed between two pieces of glass. The thickness of the liquid crystal sample usually is kept small so that light can easily pass through it. The liquid crystal sample is positioned between two polarizers, which are adjusted so that they cross each other. This arrangement would normally ensure that no light comes through to your eye; but, because of the orientational order of liquid crystals, light does reach your eye. (I will discuss polarized light at length later.) Instead, you see a collection of curved lines as shown in plate 1. The lines you see in the microscope are due to defects within the liquid crystal, which also will be discussed later. The name for the liquid crystal phase I have been discussing stems from these defect lines, which reminded some of the early scientists of threads; so the name *nematic* liquid crystal comes from the Greek word for thread.

As is often the case in science, things turn out to be more complicated than they first appear. Molecules that possess intermolecular forces and stay parallel to one another form the nematic liquid crystal phase just as I have described. Molecules with intermolecular forces that favor alignment between molecules at a slight angle to one another form a slightly different phase. In this liquid crystal phase, the director is not fixed in space as in a nematic phase; it rotates throughout the sample. A representation of this phase is shown in figure 1.6, again using elliptical shapes to represent the molecules. The molecules appear shorter at some points in the diagram because they are viewed head-on.

The best way to visualize the helical pattern formed by the director as it changes its direction is to imagine the motion of a nut as you screw it onto a bolt. As you rotate the nut around and around, the nut moves along the axis of the bolt. This is exactly what the director does in this type of liquid crystal.

Notice that it takes the director a certain distance to rotate one full turn, just as the nut moves a certain distance along the axis of the bolt as it is rotated

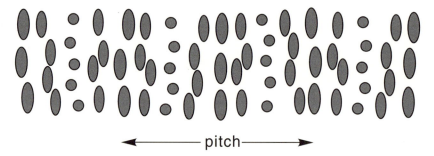

Fig. 1.6. Snapshot of the molecules in the chiral nematic liquid crystal phase. The direction of preferred orientation (director) rotates about a horizontal axis in the plane of the page. The pitch is the distance for one full director rotation.

one complete turn. This distance is called the *pitch* of the liquid crystal. In actuality, the twisted structure repeats itself over a distance equal to one-half the pitch, since the director can always be defined pointing in either direction along the preferred direction. This fact will be important when we discuss how light interacts with these twisted liquid crystals.

The most common examples of molecules forming this phase are closely associated with cholesterol, so we call them *cholesteric* liquid crystals. Cholesteryl myristate has a cholesteric liquid crystal phase. The term cholesteric is not a good one, however, since there are many cholesteric liquid crystals that have no connection with cholesterol whatsoever. A more proper name for this phase is *chiral nematic* liquid crystal (chiral simply means twisted). The twist present in chiral nematic liquid crystals produces some spectacular optical properties. I will discuss them later, but some of these properties make chiral nematic liquid crystals appear quite different under a microscope.

Plate 2 shows an example of a chiral nematic liquid crystal under the microscope; notice that defect lines are still present, but the patterns are different due to the twist inherent in the sample. The liquid crystal in plate 2 is oriented with the helical axis along the direction of viewing (perpendicular to the page). This is called the *Grandjean texture*, after the French scientist F. Grandjean, who worked with similar samples around 1920. Plate 3 shows another picture of a chiral nematic liquid crystal, but the helical axes in this sample are perpendicular to the viewing direction. The pitch of this liquid crystal is long enough to be visible under the microscope, so all the lines in the picture are points where the director rotates back to the same position. When this occurs, it is called the *fingerprint texture*.

A substance may possess either the nematic liquid crystal phase or the chiral nematic liquid crystal phase but not both. However, there is another type of liquid crystal phase that can occur as the only liquid crystal phase a substance possesses or at a temperature below the nematic or chiral nematic phase of a substance. This third liquid crystal phase is called the *smectic* phase, from the Greek word for soap. The early investigators noticed that these liquid crystal phases possessed mechanical properties reminiscent of soaps. In fact, the thick "goo" usually found in the bottom of a soap dish is a liquid crystal phase (we will discuss this type later) and has all the properties of a smectic liquid crystal. In the smectic phase, not only is the small amount of orientational order of liquid crystals present, but there is also a small amount of positional order. Again, the molecules are free to bounce around quite randomly, but in this phase they tend to point along the director and arrange themselves in layers. To be more exact, a snapshot in time would reveal that slightly more molecules tend to be positioned in regularly spaced planes with fewer molecules in between. Likewise, following a single molecule would reveal that it spent slightly more time in these planes than between these planes. Figure 1.7 illustrates this small amount of both orientational and positional order. Notice that there are two slightly different smectic phases. In the smectic A phase the director is perpendicular to the

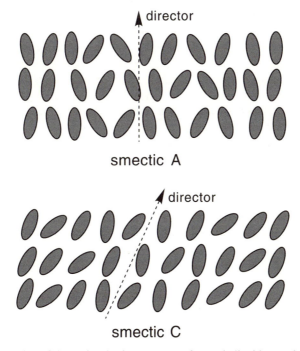

Fig. 1.7. Snapshot of the molecules in two types of smectic liquid crystal phases. The director is perpendicular to the layers in the smectic A phase but makes an angle other than 90° in the smectic C phase.

planes, while in the smectic C phase the director makes an angle other than 90° to the planes. Smectic liquid crystals have a unique appearance under the microscope; an example can be found in plate 4.

In both the smectic A and smectic C phases, the molecules randomly diffuse within each plane. No positional order exists within each plane, so in a sense the positional order is in one dimension only. However, other smectic liquid crystal phases exist in which the molecules are somewhat ordered within each plane. In other words, a molecule diffusing through the plane spends more time at special locations than at other locations. The positional order is three-dimensional now. Various arrangements of these special locations are possible, and these phases have been given names like smectic B or smectic E. The designation of smectic phases by letters is more historical than physical, in that additional letters have been used each time a new smectic phase is discovered. A recent list of these smectic phases included the letters A through K. The additional order of these smectic phases makes them appear quite different under a microscope. Plate 5 is a picture of a smectic B phase.

Although many compounds exhibit only one liquid crystal phase, it is not unusual for a single substance to possess more than one. A compound called *p*-azoxyanisole (PAA) has only a nematic phase; its phases are shown in figure 1.8. The figure also contains a schematic of the PAA molecule. In diagrams of molecules, letters are used to represent the atoms, with short straight lines representing the fact that two atoms share electrons and therefore are bound together. For simplicity, however, some of the carbon atoms are denoted by points where one or more lines (representing bonds) meet. The hydrogen atoms bound to these carbon atoms are also omitted. You can tell where they are by realizing that every carbon atom must share its four electrons. The number of hydrogen atoms bound to a carbon atom is therefore four minus the number of lines (other bonds) coming together to represent the carbon atom.

Fig. 1.8. Phase diagram for *p*-azoxyanisole (PAA). In the diagram of the molecule, intersecting lines (which represent chemical bonds) denote carbon atoms. For simplicity, the hydrogen atoms bonded to these carbon atoms have been omitted.

Fig. 1.9. Phase diagram and molecular structure of 4-*n*-pentylbenzenethio-4'-*n* decyloxybenzoate ($\overline{1}$0S5).

Fig. 1.10. Complete phase diagram for cholesteryl myristate along with its molecular structure.

A compound slightly different from PAA is called 4-*n*-pentylbenzenethio-4'-*n*-decyloxybenzoate ($\overline{1}$0S5 for short!). It possesses three liquid crystal phases, as shown in figure 1.9. Cholesteryl myristate has both a chiral nematic and a smectic A phase; its full diagram is in figure 1.10.

DISCOTIC LIQUID CRYSTALS

So far the entire discussion has been devoted to *calamitic* liquid crystals, that is, the phases formed by rodlike molecules. The reason for this is simple; these phases are the most common liquid crystal phases and therefore the most well-known. In 1977, however, researchers in India discovered that disclike molecules also form liquid crystal phases in which the axis perpendicular to the plane of the molecule tends to orient along a specific direction. These phases and the molecules that form them are called *discotic* liquid crystals.

nematic   columnar

Fig. 1.11. Illustration of the nematic and columnar discotic liquid crystal phases. The flat discs represent molecules.

The most simple discotic phase is also called the *nematic* phase, because there is orientational order but no positional order. Such a phase is shown in figure 1.11, where the molecules have been drawn as flat discs. The molecules move about quite randomly, but on average the axis perpendicular to the plane of each molecule tends to orient along a special direction called the director. As evident from the figure, this phase looks like a pile of coins. The *columnar* or *smectic* discotic phase is also shown in figure 1.11. In addition to the orientational order present in the nematic discotic phase, most of the molecules tend to position themselves in columns. In other words, at all times more molecules are located in the columns than between columns. The columns are arranged in a hexagonal lattice. A nice analogy to this phase is a set of stacked coins, with the stacks positioned in the centers and vertices of connected hexagons. However, one must be careful about such analogies. The coins in a stack have a great deal of positional order (i.e., the distance between coins is fixed and the same for all coins), whereas the molecules of a columnar or smectic discotic liquid crystal are positioned in the stack quite randomly. The positional order in this phase is therefore two-dimensional. *Chiral nematic* discotic liquid crystals also exist. In this phase the director rotates in a helical fashion throughout the sample, just as in the case of rodlike chiral nematic liquid crystals. An example of a molecule possessing discotic liquid crystal phases is shown in figure 1.12 along with a temperature diagram of these phases. Notice that this molecule has a fairly rigid, planar center with hydrocarbon chains emanating in all directions. These features are common to just about all discotic liquid crystal molecules.

Although the structure of discotic liquid crystals is quite different from that of other types of liquid crystals, their appearance under the microscope is similar. Plate 6 shows a "crystal" of a discotic liquid crystal as it forms from the isotropic liquid.

Fig. 1.12. Phase diagram and molecular structure of a typical discotic liquid crystal.

OTHER TYPES OF LIQUID CRYSTALS

One area of modern technology that also deals with liquid crystal phases is the polymer industry. *Polymers* are extremely long and slender molecules that form when chemical reactions link shorter molecules together. A bowl of cooked spaghetti is an especially good model for most polymers. Two types of polymers give rise to liquid crystal phases. The first is composed of fairly rigid segments (like the central part of a liquid crystal molecule) connected together end to end by flexible segments (like the ends of a liquid crystal molecule). Although these long polymers move around and collide with each other in the liquid crystal phase, the rigid segments tend to remain pointing in one direction. The second type of polymer is one composed of a single, completely flexible polymer with rigid segments attached as side chains along its

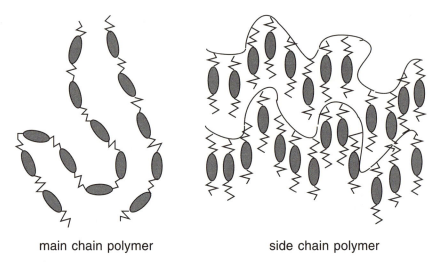

main chain polymer　　　　　　　　　side chain polymer

Fig. 1.13. Illustration of the orientational order in a main chain and a side chain liquid crystal polymer. The elliptical shapes represent rigid segments of the polymer, and the zigzag lines represent flexible segments.

length by short flexible segments. In the liquid crystal phase of this type of polymer, the long flexible part winds its way throughout the substance without any orientational or positional order, but the rigid segments attached to it exhibit the orientational order typical of liquid crystal molecules. Figure 1.13 illustrates both of these types of liquid crystal polymers. Nematic, chiral nematic, and smectic phases have been found in polymers, with some possessing more than one liquid crystal phase. Their appearance under a microscope resembles that of other liquid crystals in many ways. A photograph of a nematic polymer liquid crystal is shown in plate 7. Liquid crystal polymers are an exciting new field of modern technology, which is covered in some depth in chapter 9.

Before our introduction to the liquid crystal phase can be complete, we must consider one other class of compounds in which liquid crystal phases are important. In the discussion thus far, we have considered only pure substances, and temperature changes have caused the liquid crystal phases to come and go. We call this class of liquid crystal substances *thermotropic* liquid crystals. In some cases when two different substances are mixed together, the mixture can exhibit different phases not only as the temperature is changed but also as the concentration of one component of the mixture is varied. When the liquid crystal phase is dependent on the concentration of one component in another, we call such a system a *lyotropic* liquid crystal.

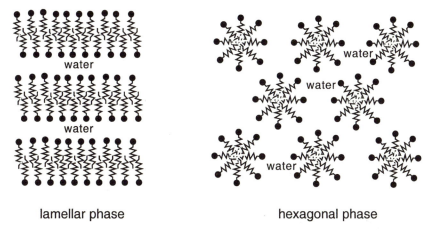

lamellar phase   hexagonal phase

Fig. 1.14. Cross-sections of the lamellar and hexagonal lyotropic liquid crystal phases. The round end of each molecule represents a hydrophilic (water-seeking) group, while the zigzag line depicts a hydrophobic (water-fearing) group.

The easiest way to put together a lyotropic liquid crystal mixture is to start with a molecule that has end groups with different properties. For example, one end of the molecule could show an affinity for water while the other end tends to exclude water. When such molecules are placed in water, the ends that exclude water tend to arrange themselves together, allowing the other end of the molecules to be in contact with the water. This effect results in structures of various shapes (spheres and cylinders are common), which themselves can be positioned in a very specific arrangement. Again the molecules are free to roam about. But as they do so, they retain the orientational and positional order due to these structures, and are therefore proper liquid crystalline phases. We can represent molecules of this type by a circle (water-seeking) connected to a tail (water-excluding). Two of the many structures these molecules form are shown in figure 1.14.

There are two important examples of molecules that behave as I have been describing. The first is soap molecules, and the ability of soap to dissolve oil and dirt is directly related to the ordered structures the soap molecules form in water. The second example is biologically important molecules called *phospholipids*. The ordered structures these molecules form are found everywhere in biological systems and include the cell membrane itself. In both of these types of molecules, one end of the molecule is charged and tends to associate with water while the other end is composed mostly of carbon and hydrogen, which tend to exclude water. Lyotropic liquid crystals are important not only for their biological significance but also because of their use as surfactants. Chapter 8 addresses both of these areas.

Chapter 2

THE STORY OF LIQUID CRYSTALS

FOR ONE WEEK during the summer of 2000, nearly eight hundred scientists from forty-four countries gathered in Sendai, Japan, for the eighteenth International Liquid Crystal Conference. The number of researchers participating in this conference testified to the importance of liquid crystal research and technology in today's world. More than one presentation at this conference took a historical perspective on the science of liquid crystals since their "discovery" in 1888. The story of these 112 years, how interest in liquid crystals waxed and waned, and how the field blossomed thirty years ago, is truly a fascinating one.

BEFORE THE "DISCOVERY"

Although individual scientists are frequently given credit for important scientific discoveries, in many cases there are other scientists who also deserve a great deal of credit. Usually these people perform experiments or introduce new ideas that are very similar to those of the "discoverer." Yet these workers for one reason or another do not put their results or ideas together in a way that fully captures the significance of their new work. Sometimes the "discoverer" is the scientist who recognizes the meaning of this new work, performs a few experiments that show the true meaning in an unambiguous way, and then clearly explains it to the rest of the world. It is therefore not surprising to hear comments like "I was in the right place at the right time" coming from these "discoverers."

The story of liquid crystals starts with a few European investigators who observed some new and interesting phenomena but who never fully realized exactly what was happening in their experiments. These early scientists are not given the credit for discovering liquid crystals, although we are now sure they observed liquid crystal phases in their laboratories. In fact, their names are often omitted in historical accounts.

The time frame of this early work is 1850 to 1888, and three different types of experiments were involved. The first type concerned the study of some biological specimens using a microscope. It was noticed that the outer covering of a nerve fiber formed soft and flowing forms when left in water. It was also observed that these forms produced unusual effects when polarized light was used. One European scientist, Rudolf Virchow, left a report of his experiments in his archives; the German ophthalmologist C. Mettenheimer reported his observations in a scientific publication; and a short monograph by G. Valentin that summarized the results of polarization microscopy on biological materials was published in Leipzig. It is easy to understand why these scientists found their results interesting; liquids produce no unusual effects when polarized light is used, whereas solids do. The biological specimens were not solid, and yet the unusual polarization effects were present. Further progress using these specimens was probably hindered by the fact that the biological specimens were complicated structures, thus making analysis of the phenomenon quite difficult.

The second type of early experiment involved the study of how substances crystallize. In Karlsruhe, the German physicist Otto Lehmann constructed a heating stage for his microscope, allowing him to observe how materials crystallized as he slowly lowered the temperature of his samples. Although at this time biologists and physicians were using polarized light to view biological samples, Lehmann was one of the first physical scientists to use a polarizing microscope when he later added polarizers to his heating stage microscope. This instrument was a significant technical advancement that later allowed Lehmann to be one of the most important early contributors to the field of liquid crystals. A sketch of his apparatus appears in figure 2.1.

Fig. 2.1. Otto Lehmann's microscope and heating stage. A flame from the burner (under the stage) heated the stage while air from the tube (above the stage) cooled it. Careful adjustment of these two devices allowed Lehmann to precisely control the temperature of his samples.

The success of Lehmann's microscope in furthering scientific knowledge is an excellent example of how a technological achievement can dramatically affect the course of science. With this new apparatus, scientists could control the environment of their samples and observe their behavior under a microscope. This instrument allowed them to perform experiments that were previously impossible and observe phenomena that earlier were hidden from the investigators' eyes. Liquid crystal research would never be the same. In fact, a polarizing microscope with a precision heating stage is a standard piece of modern equipment; today no liquid crystal laboratory works without one.

Lehmann noticed that some substances did not crystallize from a clear liquid but changed instead to an amorphous form, which then crystallized. He did not realize at this point that a new phase of matter was being observed; he conjectured that it was somehow related to the transition from the liquid phase to the solid phase. It is easy to understand why Lehmann might have thought in this way. Phase transitions do not always occur in the idealized way. For example, the presence of impurities causes phase transitions to occur over a narrow temperature range where both phases coexist. Certainly Lehmann had observed many phase transitions under the microscope and knew they varied significantly from one substance to the next. The amorphous form he saw in his microscope might have struck him as just another less than ideal solid to liquid phase transition.

The third type of experiment performed in this time period was one with compounds synthesized from cholesterol. P. Planer in the city of Lvov (now part of the Ukraine), the German chemist W. Lobisch, and B. Raymann in Paris all reported that these compounds displayed striking colors when cooled. Although it certainly was an unusual phenomenon, none of these people had any idea that these colors were coming from a phase of matter other than the solid or liquid phase. One other investigation from this time period deserves to be mentioned. Around 1850 the chemist W. Heintz, who was studying natural fats, reported that stearin possesses an unusual melting behavior. It turns cloudy around 52°C, is completely opaque by 58°C, and becomes clear at 62.5°C. He and a colleague at one point referred to the change at 62.5°C as a second melting point. As will be evident in the next section, these observations were extremely similar to the experiments that forty years later led directly to the "discovery" of liquid crystals.

THE "DISCOVERY"

The person generally given credit for "discovering" liquid crystals is an Austrian botanist named Friedrich Reinitzer. Reinitzer's main interest was the function of cholesterol in plants. Of course, the structure of cholesterol was not known at the time. In 1888, when Reinitzer observed the melting behavior of an organic substance related to cholesterol, he also described it by saying that it possessed two melting points. At 145.5°C it melted to a cloudy liquid, and at 178.5°C this cloudy liquid turned into a clear liquid. He also described some of the same color phenomena that had been reported by earlier scientists working with cholesterol derivatives: a blue color briefly appears upon cooling when the clear liquid turns cloudy, and a blue-violet color occurs just before the cloudy liquid crystallizes.

Why Reinitzer stated that the substance seemed to have two melting points, while earlier scientists frequently associated unusual behavior of this sort with the crystallization process, is one of those intriguing questions that often arise in the history of science. Was he aware of the work of Heintz? We cannot be certain. What we do know is that twenty years later he mentioned Heintz's experiments in a letter to Otto Lehmann. Regardless of whether or not Reinitzer was the first to describe a substance as having two melting points, his description of the cholesterol derivative opened a line of inquiry that turned out to be extremely fruitful. If the solid phase melted at 145.5°C, what was melting at 178.5°C? Might this be a form of matter other than the solid or liquid phase? Because he described what he saw in a way that posed the right questions for future investigators, Reinitzer is usually called the "discoverer" of liquid crystals.

The substance Reinitzer observed was in fact cholesteryl benzoate, a chiral nematic liquid crystal that has been the object of many investigations over the years. Reinitzer made this compound from cholesterol, a natural substance found in both plants and animals. Interestingly enough, the blue color Reinitzer saw just as the sample turned from clear to cloudy became a very important problem for condensed matter physicists around 1980; but a discussion of this must wait until a later chapter.

Reinitzer knew of the work of Lehmann and saw the connection between his own observations and Lehmann's findings. He therefore sent some of these samples to Otto Lehmann, who at this time was a professor of natural philosophy (physics) in Germany. Lehmann performed many experiments on this substance with his heating stage microscope. In one report he included a drawing that is extremely similar to the picture of a chiral nematic shown in plate 2. Lehmann described Reinitzer's substance along with substances be received from other investigators in a variety of ways during this time period. First he called them soft crystals that were almost fluid; then the term "floating crystals" was used, which was later replaced with "crystalline fluids." He gradually became convinced that the cloudy liquid was a uniform fluid phase, but one that affected polarized light in a manner typical of solid crystals, not liquids. This combination of characteristics— flow properties like a liquid and optical properties like a solid—finally led

Lehmann to label these substances "liquid crystals." In spite of many subsequent arguments over nomenclature, this was the name that eventually survived.

I must make note of the symbolism in Reinitzer's sending of his samples to Lehmann. Reinitzer was an Austrian botanist and Lehmann was a German physicist, so the act represented a collaboration between scientists representing both two scientific disciplines and two different countries. Thus, from the beginning, liquid crystal research was an international and multidisciplinary field, a fact that has grown more true with each passing year. Chemists, physicists, biologists, engineers, and medical doctors from almost all of the industrialized countries of the world are involved in liquid crystal research. They attend the same scientific meetings and spend time as visiting scientists at each other's laboratories. The spirit of international and interdisciplinary collaboration displayed by these two early liquid crystal researchers is still the hallmark of the field.

AFTER THE "DISCOVERY"

Lehmann remained the dominant figure in liquid crystal research around the turn of the century. He experimented with the first nematic liquid crystal, which was synthesized by two German chemists, Ludwig Gattermann and A. Ritschke. This compound was also the first liquid crystal not based on a natural substance. In fact, Gattermann and Ritschke supplied him with a series of substances, and Lehmann noticed that some liquid crystals behaved differently from others. He used different words to describe them but did not choose to call them two different types of liquid crystals (now known as nematic and smectic). Lehmann also observed that a solid surface in contact with a liquid crystalline substance causes the liquid crystal to orient in a certain direction. This idea was of paramount importance when modern scientists began to experiment with liquid crystal displays.

Not all of Lehmann's ideas were accepted by other investigators. Gustav Tammann, a solid-state chemist, and Walter Nerst, a physicist, argued that the unusual phenomena could be explained by the fact that the substance was a mixture or emulsion of two distinct compounds or phases. Lehmann used the results of his own experiments, together with those of the physical chemist Rudolf Schenk, to argue more and more convincingly that a single phase was involved.

Another important contributor at this time was the German chemist Daniel Vorlander, who worked in Halle. Workers in his institute synthesized many new liquid crystalline substances and were the first to observe a single substance that possessed more than one liquid crystal phase. Out of this work Vorlander was able to identify what kinds of substances were likely to be liquid crystalline. Being able to make this identification was of obvious practical importance, but his suggestion that a linear molecular shape was important influenced both theoretical and experimental work for many years to come. As time went on, more results about new liquid crystals poured out of Vorlander's group, with over eighty doctoral theses being written on liquid crystals between 1901 and 1934. Today the Martin Luther University in Halle still boasts an important liquid crystal institute. In fact, there has been continuous liquid crystal research at Halle from about 1900 to the present.

Vorlander's suggestions as to what kinds of substances were likely to have liquid crystalline properties were a strong influence on the theoretical and experimental work of the physicist Emil Bose. Bose attempted to construct a complete theory of liquid crystals based on molecular structure. This was an important contribution, because up to this point there had been considerable confusion on Lehmann's part as to what the ordering unit was and what changes (chemical and/or physical) took place at the transition. Bose's theoretical work also represented a strong argument against the hypothesis that liquid crystals result from some sort of emulsion. A little less than ten years later, Max Born suggested that the basic interaction between the linear molecules that causes the liquid crystal phase is due to a slight separation of positive and negative charges on the molecules.

Lehmann himself was partially responsible for introducing the study of liquid crystals in France. In 1909 he was invited to a conference in Paris. The culmination of the subsequent work in France was the publication in 1922 of a paper by Georges Freidel describing the different liquid crystal phases. In this paper Freidel proposed the classification scheme using the words nematic, smectic, and cholesteric. The accepted description of the liquid crystal phase now involved the idea of molecular ordering, laying to rest numerous other suggestions as to the origin of liquid crystallinity. He also explained that the lines one sees in liquid crystal phases under the microscope were defect structures, which represent drastic changes in the direction of orientation. Through careful analysis of the these defect structures, Freidel correctly deduced that smectic liquid crystals possess a layered structure. Additionally, he clearly understood that a liquid crystal could be oriented by an electric field. The effect of electric and magnetic fields on liquid crystals later became the subject of great attention.

The period between 1922 and World War II was characterized by advancement in several different areas by scientists in many countries. Theoretical work into the elastic properties of liquid crystals grew out of the work of Carl Oseen in Sweden and reached fruition in the "continuum theory" of F. C. Frank in England. Using these new theoretical ideas, these scientists explained why liquid crystals adopted various orientational configurations. X-ray experiments in France and Germany revealed in a most unambiguous way that liquid crystals possess more order than liquids but less than solids. The action of electric and magnetic fields on liquid crystals began to be understood, mainly due to experiments performed in Holland and the former Soviet Union. Finally, some progress was made in understanding the light-scattering properties of liquid crystals, explaining for the first time why liquid crystals appear cloudy.

Progress was slow during the war years, but two new areas of success should be pointed out. First, the flow properties of liquid crystals in high electric and magnetic fields were investigated. Second, the "degree of order" was described by the order parameter

$$S = \text{average of } \left(\tfrac{3}{2}\cos^2\theta - \tfrac{1}{2}\right).$$

Choosing to use this order parameter may seem like a fairly unimportant step, but the opposite turned out to be true. We now know from a combination of both theoretical and experimental results that a full description of the orientational order present in liquid crystals is quite a complicated task. Fortunately, the above order parameter comes closer to describing the actual orientation order than any other single parameter. The scientists of the 1940s suspected this result to be the case, and nearly all future work was based on descriptions utilizing this order parameter.

Recent Developments

Interest in liquid crystals all but disappeared after World War II. In describing a possible reason for this lack of interest, F. C. Frank remarked in 1958 that too many people felt that all the important problems concerning liquid crystals had been solved. Another reason given is that writers of chemistry and physics textbooks during this period did not discuss the liquid crystal phase, so many scientists did not know that certain unusual melting phenomena involved a completely different phase of matter. James Fergason, one of the pioneers in the development of liquid crystal displays, felt that the most important force behind this period of quiescence was the apparent lack of any practical applications for liquid crystals.

Whatever the reason for the lack of interest immediately after World War II, the situation began to change shortly before 1960. At this time a few individuals undertook a general reexamination of liquid crystals in the hope of learning more about their molecular structure, optical properties, and technical possibilities. In 1957 Glenn Brown, an American chemist, published an article reviewing the liquid crystal phase. In 1958 the Faraday Society of London held a conference on liquid crystals. At about the same time, scientists at the Westinghouse Research Laboratories began a research project focused on liquid crystals. A group led by I. G. Chistiakoff began work in the former Soviet Union; and in 1962 George Gray, an English chemist, published a full-length book describing the molecular structure and properties of liquid crystals. Brown was instrumental in founding the Liquid Crystal Institute at Kent State University and also organized the first of a series of international liquid crystal conferences that continues to this day. The conference in Sendai was part of this series.

Immediate progress was made in understanding the molecular structure of liquid-crystal-forming compounds. Two German physicists, Wilhelm Maier and Alfred Saupe, formulated a microscopic theory for the liquid crystal phase that did not hypothesize a separation of charge on the molecules. The importance of this last result should be emphasized. For the first time, liquid crystal researchers possessed a theory that predicted the behavior of the phase starting with realistic characteristics of the individual molecules. Also, it quickly became apparent that liquid crystalline substances had the ability to detect extremely small changes in temperature, mechanical stress, electromagnetic radiation, and chemical environment; the door was thus opened to a number of possible applications. In 1968 two scientists at RCA demonstrated that a thin layer of liquid crystal was capable of switching from cloudy to clear when an electrical voltage was applied. Although this was the first liquid crystal display (LCD), it required too high a voltage, consumed too much power, and produced a display of poor quality. Three years later, two scientists working in Switzerland and one in the United States reported that a liquid crystal cell could be turned from clear to black by an electrical voltage. The improved quality of this display together with its low power consumption made this device worthy of application. Within ten years liquid crystal displays that required extremely little power were being used by manufacturers of battery-operated equipment such as watches and calculators.

Perhaps the most important advance during this period was the synthesis of the first moderately stable room temperature liquid crystal, *p*-methoxybenzylidene-*p*-*n*-butylaniline (MBBA). The phases of MBBA along with its molecular structure are shown in figure 2.2.

Fig. 2.2. Phase diagram and molecular structure of *p*-methoxybenzylidene-*p*-*n*-butyl aniline (MBBA), the first fairly stable room temperature liquid crystal.

The combination of both basic scientific progress and new technological ideas caused an explosion in the number of researchers in the field. In the area of LCDs alone, at least six different types of displays were developed, including some that used chiral nematic and smectic liquid crystals. Research papers began to fill volumes each year, and new applications appeared regularly. Research conferences on liquid crystals grew in size from roughly fifty participants in the early 1960s to nearly seven hundred people in the late 1980s. Groups of physicists, chemists, and engineers studying liquid crystals formed at major universities and industrial laboratories throughout the world. This period was clearly the most exciting time in the history of liquid crystal research, and much of our detailed knowledge about liquid crystals stems from work done at this time.

During this period of time, the study of liquid crystals played an important role in our increasing understanding of how molecules behave cooperatively and how molecular structure influences this behavior. Chemists synthesizing new liquid crystal compounds contributed to our understanding of organic synthesis in general. Liquid crystals were used as a solvent or medium in which to probe other substances. This flurry of activity also produced liquid crystal displays for watches, calculators, clocks, telephones, cameras, office equipment, personal computers, miniature TVs, automobile dashboards, and windows switchable between clear and opaque. Other applications included liquid crystal thermometers and temperature-sensing films, high-strength liquid crystal polymers, and surfactants for the oil recovery industry. Progress in our understanding of the liquid crystal phase also aided our understanding of the cell membrane and of certain diseases, such as sickle-cell anemia and arteriosclerosis. Many of these important contributions will be discussed fully in later chapters.

The last ten years have shown a maturing of the field of liquid crystals. Researchers are asking more specific questions about the behavior of liquid crystals and using more sophisticated equipment to obtain much more detailed information on molecular conformations, phase structures, and liquid crystalline properties. In many cases the new information serves to clarify prior conceptions; in some cases old ideas must be discarded. One area where discoveries continue to be made is in enumerating the number of possible ways liquid crystal molecules organize themselves into condensed phases. Anti-ferroelectric ordering in chiral smectic phases, the unique way "banana-shaped" molecules form ordered phases, and the many ways defects in the ordering of the liquid crystal molecules spontaneously stabilize structures testify to the ingenious organizational skills of nature. The study of the interactions between liquid crystals and surfaces has revealed a great deal of new information, opening the door to both improved and new applications. Computer simulations of liquid crystal phases have achieved much more sophistication, successfully describing the isotropic, nematic, smectic, and crystal phases of actual molecules. Such simulations are now used to explore the phase behavior of yet-to-be-synthesized compounds to help direct synthesis efforts in fruitful directions.

During this same period, the study of liquid crystals has crossed into related areas. Mixtures of liquid crystals and polymers have properties that may make them extremely useful for applications. The study of liquid crystals dissolved in isotropic liquids and isotropic liquids dissolved in liquid crystals has opened several lines of investigation shedding new information about condensed phases and their interactions with one another. The self-organization of biologically important molecules and the production of self-assembled layers on substrates are two more examples of where liquid crystal science has made important contributions. While there have been some new applications of liquid crystals recently (flat panel computer displays and electronic books, for example), what has characterized the last ten years has been the increase in performance and reliability over earlier liquid crystal applications. One need only look at current laptop computer screens to see just how much they have improved. There seems to be no end in our progress to understand liquid crystals, so we can be certain that just as many new developments will be made by scientists and engineers of the future.

Chapter 3

ELECTRIC AND MAGNETIC FIELD EFFECTS

THE RESPONSE OF liquid crystals to electric and magnetic fields is perhaps the best illustration of the delicate nature of this phase of matter. Solids, liquids, and gases respond to electric and magnetic fields, but the response is minimal even when strong fields are applied. Liquid crystals, on the contrary, respond to even weak electric and magnetic fields with significant structural changes. This chapter explores these effects by first describing the reasons behind this delicate response to electric and magnetic fields and then discussing the many different ways in which this response can take place.

ANISOTROPY

Although liquid crystals are fluids, the fact that the molecules on average orient themselves along a certain direction in space has a profound effect on the properties of the phase. The best way to understand this is to consider the case of liquids first. In the liquid phase (or the gas phase) the completely disordered motion of the molecules produces a phase in which all directions in space are equivalent. In other words, any property measured along one direction of space has the same value when measured along any other direction. For example, suppose sound is made to propagate through a liquid. If we measure how the sound loses its intensity as it passes through the liquid, we obtain the same answer no matter in which direction the sound is traveling. The property of obtaining the same result regardless of direction is called *isotropy*, and a phase that has this property is called an *isotropic* phase. All liquids and gases are isotropic.

If one imagines our picture of the liquid crystal phase, it is not difficult to realize that this phase is not isotropic. At every point in a liquid crystal the molecules define a special direction by spending more time pointing along this direction than any other direction. Sound traveling along the director of a liquid crystal encounters molecules that tend to lie along the direction of propagation, while sound traveling in other directions encounters molecules oriented on average at an angle to the direction of propagation. Since a group of molecules responds differently to sound according to the direction the sound is traveling relative to the long axis of the molecules, sound traveling along the director will lose its intensity in a different way than sound traveling in any other direction. How the sound loses its intensity in these two cases depends on the actual molecule that makes up the liquid crystal. Typically, sound loses its intensity over a shorter distance when traveling along the director. This property is called *anisotropy* (no isotropy), and accordingly the liquid crystal phase is called an *anisotropic* (not isotropic) phase.

For most people, the difference between a solid and a liquid is obvious. Solids are "rigid," and liquids flow. As simple as this distinction is, there are some substances that are not easily placed in one category or the other. For example, some plastics flow so slowly that an observation of a plastic sample over a short time interval might lead to the conclusion that does not flow. Classifying it as a solid, however, would be misleading. Anisotropy is another property used to distinguish solids from liquids. Since no liquid is anisotropic, any anisotropic substance must be a solid (unless it is a liquid crystal). The correspondence between anisotropy and solids is not perfect, however, because some solids are isotropic. By using several criteria to classify whether a substance is a liquid or a solid, scientists can deal more successfully with those substances which do not behave in the normal manner. Anisotropy is the most important characteristic shared by both liquid crystals and solids. Otto Lehmann recognized this fact when he noted that light interacted with this "new substance" just as it did with anisotropic solids. No liquid was capable of such behavior.

This anisotropy manifests itself in many physical properties. When external fields (electric or magnetic) are applied to a liquid crystal, the phase responds differently depending on whether the field is applied along the director or at an angle to it. Since light is an electromagnetic wave, how light propagates through a liquid crystal depends on its direction relative to the director. A very different example concerns the mechanical properties of liquid crystals. Imagine placing some liquid crystal between two flat plates and measuring the force necessary to slide one plate over the other. As shown in figure 3.1,

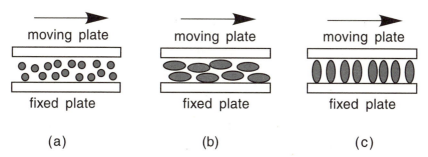

Fig. 3.1. Testing anisotropy in a liquid crystal. The force necessary to slide the top plate over the bottom plate with liquid crystal in between depends on the orientation of the director. The necessary force is less when the director points out of the page (a) or along the direction of plate movement (b) than when the director points up or down (c).

the director of the liquid crystal can point in three different directions relative to the plates and direction of movement. The force necessary to slide the plates over each other is different for each case, being nearly the same for cases (a) and (b) but much higher for case (c). No such effect exists for a liquid placed between the two plates. These results are not difficult to reconcile if one realizes that this situation is similar to what happens when logs float down a narrow river. If the logs point downstream, they flow without the problems encountered if they flow pointing across the river. These two cases are analogous to (b) and (c) in the liquid crystal, respectively. The analogy to case (a) in the liquid crystal is not possible, since logs floating in water rarely orient themselves vertically!

ELECTRIC FIELDS

Electricity is the phenomenon associated with charged objects. As we all know, the charge on such objects comes in two varieties (positive and negative), with opposite charges attracting each other and like charges repelling each other. The two charged objects need not be in contact to experience this force; two charged objects separated by any distance attract or repel each other, although the strength of the force does decrease as the distance between the two objects increases. Scientists use the idea of an *electric field* to describe the fact that charged objects experience a force due to other charged objects even though there may be some distance between them.

The idea of an electric field is a simple one. Consider a single charged object sitting at a point in space. If a second charged object is placed near it, an electric force between the two objects occurs. The presence of the first charged object somehow alters the space around it, since the second charged object behaves differently if placed in this space depending on whether the first object is present or not. To describe the fact that the space is altered, we say that the first charged object produces an electric field around it. When a second charged object is placed in this electric field, a force is produced. In our minds, therefore, we have divided this phenomenon into two stages: the first charged object produces an electric field, which in turn produces a force on the second charged object. The idea of an electric field might appear to be quite an artificial construction, but under some circumstances the electric field continues to exist after the charged object producing it has been removed. One therefore might conclude that the idea of the electric field has some validity beyond that of helping us visualize the electric force. In fact, this approach (associating a field with a force) has been extremely useful over the last one hundred years, successfully describing not only the electrical force but all the fundamental forces of nature. It should be noted that the electric field from a single charged object extends to infinity, since the electric force weakens with distance but does not become zero. All of us thus live in the electric field produced by charged objects throughout the universe.

The electric field at any point in space has a direction associated with it. By convention, it is the direction of the force experienced by a positively charged "test" object placed there. If lines are drawn parallel to the direction of the electric field, one obtains a schematic drawing that depicts the electric field. One such diagram is shown in figure 3.2. The direction of the electric field is given by the dashed lines with arrowheads on them. The direction of the force experienced by a positive charge at point A and a negative charge at point B are given by the two solid lines with larger arrowheads. Thus electric fields can be produced in a region of space by bringing charged objects nearby.

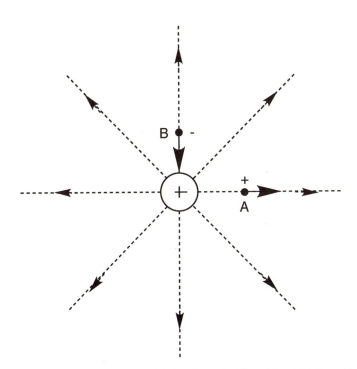

Fig. 3.2. Electric field around a positively charged object. The field lines (dashed) extend outward. The force (arrows) on nearby charged objects lies along the field lines, outward for positively (A) and inward for negatively (B) charged nearby objects.

Molecules in Electric Fields

When electric fields are applied to molecules, different kinds of molecules experience different forces. If the molecule is charged, it experiences a force in the direction of the field if it is positively charged and in the direction opposite to the field if it is negatively charged. This force tends to move the molecule along one of these two directions.

Most liquid crystal molecules are composed of neutral atoms and therefore are not charged. However, sometimes the bonding between the atoms of the molecule causes one part of the molecule to be slightly positive and another part to be equally slightly negative. This slight separation of positive and negative charge is called a *permanent electric dipole*. If no electric field is present, the permanent electric dipoles on the liquid crystal molecules are not aligned, even if orientational order is present. This is true for molecules with a slight separation of charge along the molecule because a molecule can tend to orient along the director with either the positively charged end or the negatively charged end pointing in the same direction. In the absence of an electric field, there are equal numbers of molecules with positively and negatively charged ends pointing along any direction. Likewise, a permanent electric dipole across the molecule is equally likely to orient in any direction perpendicular to the director.

The situation is much different, however, if an electric field is applied. As shown in figure 3.3, the charged parts of the molecule experience opposite forces. The two forces on the molecule do not cause the molecule to move in one direction; rather, they tend to cause the molecule to rotate until the positive and negative parts line up with the electric field. Notice that this tends to cause the molecule to orient parallel or perpendicular to the electric

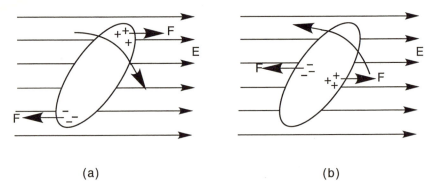

(a)   (b)

Fig. 3.3. Orientation of an electric dipole by an electric field. In (a) the dipole is along the long axis of the molecule, while in (b) it lies across the long axis. The electric field (E) causes forces (F) at the ends of the dipoles and rotation of the molecule as shown by the curved arrows.

field, depending on how the separation of charge occurs in the molecule. As the strength of the electric field is increased, the forces tending to orient the molecule along a certain direction increase.

In many liquid crystal molecules, the bonding of the atoms does not produce any separation of charge. These molecules respond to electric fields in a third way. Since the electric field produces forces on all of the atomic charges (positive nuclei and negative electrons), it often is possible for the electric field to displace the positive charges slightly in one direction and the negative charges in the other direction. This produces an electric dipole, but one which is present only in an electric field. Such dipoles are called *induced electric dipoles* and usually are much weaker than permanent electric dipoles. However, an induced electric dipole experiences the same forces in an electric field as a permanent dipole and thus tends to orient itself with the positive and negative parts of the molecule along the field.

In general, liquid crystal molecules can possess permanent or induced electric dipoles both along and across the long axis of the molecule. The molecule will orient so the larger of the two electric dipoles lies along the electric field.

Liquid Crystals in Electric Fields

Knowing how molecules respond to electric fields, it is not difficult to understand how liquid crystals behave. If the molecule is one that tends to orient its long axis along the electric field, then an electric field causes the molecules of the liquid crystal to tend to lie along the field. The orientational order is no greater than in the absence of an electric field; the difference is that the electric field causes the director of the liquid crystal to orient along the field. If the molecule tends to orient perpendicular to the electric field, then the presence of an electric field causes the director to lie perpendicular to the field. The strength of the electric field necessary to orient the director of a liquid crystal is relatively low since the director of a liquid crystal is usually free to orient in any direction. It is interesting to contrast this behavior to what happens in solids and liquids. It certainly is true that molecules with permanent or induced electric dipoles tend to orient along the electric field direction regardless of the phase they are in. In liquids, however, the disordered motion of the molecules overcomes any chance for them to orient. In a sense there is nothing in the phase that can orient. In solids the molecules are prevented from orienting with the field since the bonds between molecules prevent any change of orientation. The freedom of liquid crystal molecules to change orientation (like liquids) but to do so while maintaining some orientational order among the molecules (like solids) produces this delicate response to electric fields.

There is another way to view the response of liquid crystals to electric fields. Consider a sample of liquid crystal with its director locked in place relative to its container (later we will discuss ways in which this can be done). If an electric field is applied along the director, electric dipoles along the long axis of the molecules add together to make the liquid crystal sample itself be one large electric dipole. This fact is easily visualized. Consider a collection of rodlike molecules all pointing more or less up and down, each with the positively charged end up and the negatively charged end down. In the interior of the sample, the effect of the positively charged ends tend to be canceled by the negatively charged ends. At the top of the sample, however, there are more positively charged ends than negatively charged ends, so the top of the sample is positively charged. The same thing occurs in reverse at the bottom of the sample, where an overall negative charge occurs. The positive charge at the top of the sample and the negative charge at the bottom of the sample cause the liquid crystal sample itself to be one large electric dipole. Because the strength of the electric dipole depends on the size of the sample, scientists usually refer to the electric dipole per unit volume, since this quantity is independent of sample size. The electric dipole per unit volume is called the *electric polarization.*

If an electric field is applied perpendicular to the director (remember the director is fixed relative to its container), electric dipoles perpendicular to the long axis of the molecules add together to give an electric polarization along the field and perpendicular to the director. For each liquid crystal, the electric polarization along the director will be different from the electric polarization perpendicular to the director, even if the electric field used to produce them is the same (another example of anisotropy). Which electric polarization is larger depends on which electric dipole (along or across the long axis of the molecule) is larger. As we have seen, if the director is not constrained by other forces, the larger electric polarization (corresponding to the larger molecular electric dipole) tends to orient more, causing the director to point in a certain direction. The larger the difference in the two electric polarizations (in other words, the larger the anisotropy), the smaller the electric field necessary to orient the liquid crystal.

The larger the electric field applied to the liquid crystal, the larger the polarization. In many cases, the ratio of the polarization to the strength of the electric field is constant, but different depending on whether the electric field is applied parallel or perpendicular to the director (which is fixed relative to the container). This ratio is called the *electric susceptibility* and is a measure of how easily a material is polarized by an electric field. A graph of the electric susceptibility for a typical liquid crystal is shown in figure 3.4. Notice that the difference between the two electric susceptibilities decreases with increasing temperature. This is due to the decrease in the amount of orientational order as the temperature increases (see figure 1.5).

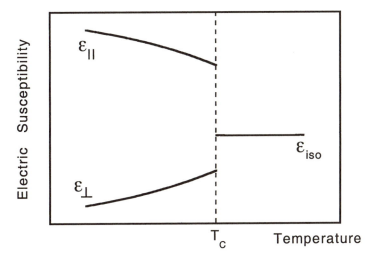

Fig. 3.4. Electric susceptibility parallel $(\varepsilon_{\parallel})$ and perpendicular $(\varepsilon_{\perp})$ to the director for a typical liquid crystal. The dashed vertical line at T_C indicates the nematic liquid crystal to isotropic liquid phase transition.

MAGNETIC FIELDS

Understanding the forces between charges becomes more complicated when the charges are moving. Not only is there an electric force between two moving charges; the motion of the charges produces an additional force called the magnetic force. Just as in the case of the electric force, two moving charges experience the magnetic force no matter what the distance between them is, with the strength of the force decreasing as the distance increases. To describe how this force works over large distances, scientists again use the concept of a field. A moving charge produces a *magnetic field* around it; when another moving charge is brought into this magnetic field, it experiences a magnetic force. Again, this concept of a magnetic field does seem to describe nature quite closely, as magnetic fields can continue to exist "on their own" after the moving charges that produced them no longer exist.

Electricity and magnetism are intimately linked, as you might guess from the fact that charges produce both effects. The two phenomena are very similar. Just as there are two types of charges that produce opposite electric effects, opposite magnetic effects are produced by a positive charge moving in one direction and a positive charge moving in the opposite direction. Furthermore, a positive charge going in one direction and a negative charge going in the same direction also produce opposite magnetic effects. Thus there are only two types of moving charges as far as magnetism is concerned, just as is true for electricity.

One important difference between electricity and magnetism can now be appreciated. To produce magnetic fields which exist for some period of time, charges must move for a similar period of time. To keep charges moving requires that they travel around a closed loop of some kind, constantly receiving energy from a device in the "circuit" in order to keep moving. It is therefore impossible to construct what would be the theoretically simplest magnetic situation, that of charge moving in only one direction. Instead, one must consider the magnetic effects produced by a small loop of moving charge as the starting point. A diagram of the magnetic field for such a loop in shown in figure 3.5. Notice that the lines that represent the magnetic field

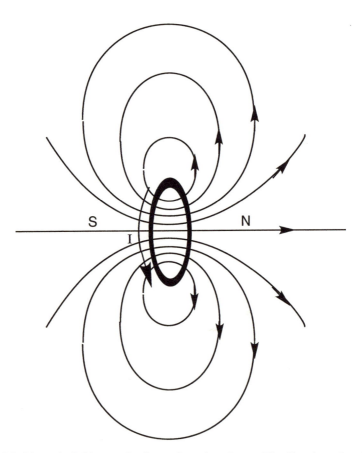

Fig. 3.5. Magnetic field around a loop of moving charge. The direction of positive charge movement is given by the arrow labeled I. The north and south poles are indicated by the letters N and S, respectively.

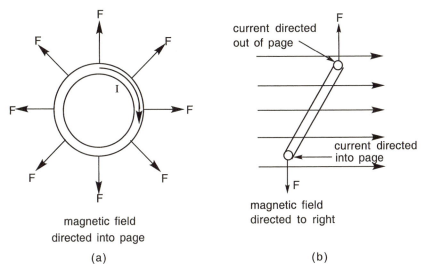

Fig. 3.6. Forces on a loop of moving charge in a magnetic field. The direction of positive charge movement is given by the arrow labeled I. With the loop broadside to the field (a), the forces on the loop (F) cancel each other and produce no net force. With the loop not broadside to the field (b), the forces do not cancel. The resulting net force causes the loop to orient broadside to the field.

are quite different from the electric field lines of figure 3.2. To help describe the magnetic field for a loop, scientists label the region to the right of the loop (where the field lines diverge from the loop) the north pole of the loop, and the region to the left of the loop (where the field lines converge into the loop) the south pole. Ordinary bar magnets act like a series of loops placed next to each other; the magnetic field lines emerge from one end of the set of loops (the north pole) and converge into the other end of the set of loops (the south pole).

The fact that the basic "unit" of magnetism is charge flowing around a loop makes the basic interaction of magnetism different from electricity. A loop of moving charge placed in a magnetic field does not experience a net force, because the loop has charge flowing in many directions. There are forces on different parts of the loop, but they add up to zero for the loop as a whole, as shown in figure 3.6(a). However, if the loop is not facing "broadside" to the magnetic field, these forces on different parts of the loop tend to orient the loop so it is "broadside." This effect is shown in figure 3.6(b). This situation is analogous to what happens to an electric dipole in an electric field. In fact, a small loop of moving charge is sometimes called a *magnetic dipole*.

Molecules and Liquid Crystals in Magnetic Fields

Although we have been using large loops of moving charge as our example, it is important to understand that nature provides us with loops of moving charge on the atomic scale. The negatively charged electrons around the nucleus of an atom sometimes behave as tiny loops of moving charge. Some atoms act as *permanent magnetic dipoles*, in effect loops of continually moving charge. These atoms are the ones that can be used to make permanent magnets. The atoms ordinarily in liquid crystal molecules are not of this type, but just as in the electric case, a magnetic field applied to the molecule causes some of the charges within the molecule to act as tiny loops of moving charge. Once this happens, the molecule possesses an *induced magnetic dipole*, and this dipole will tend to orient with its north and south poles along the magnetic field. The atoms present and how they are bound together in a molecule cause the induced magnetic dipoles to differ from one molecule to the next. Induced magnetic dipoles both parallel and perpendicular to the long axis of the molecule are possible. Liquid crystal molecules therefore tend to align either parallel or perpendicular to the magnetic field.

Just as in the electric case, the presence of a magnetic field causes a liquid crystal sample itself to possess a magnetic dipole. The magnetic dipole per unit volume of liquid crystal is called its *magnetization*. In a fashion similar to the electric case, a magnetic field applied parallel to the director (assumed locked in position relative to its container) induces a certain amount of magnetization, while a magnetic field applied perpendicular to the director induces a different amount of magnetization. This is simply another example of anisotropy in liquid crystals, which originates in the different magnetic dipoles induced along or across the long axis of the molecule. If the director is not constrained by other forces, the liquid crystal will tend to orient so the higher magnetization (corresponding to the larger molecular magnetic dipole) is parallel to the field. Again, since the director of a liquid crystal is normally not locked in place relative to its container, the director can easily change its orientation. Liquid crystals are therefore very sensitive to magnetic fields, achieving complete alignment of the director for magnetic fields of relatively low strength. The higher the anisotropy of the magnetization in the liquid crystal, the smaller the magnetic field necessary to align the director.

Just as in the case of an electric field, a higher magnetic field strength produces a higher magnetization. The ratio of the magnetization to the strength of the applied magnetic field is usually constant and is called the *magnetic susceptibility*. Liquid crystals have two magnetic susceptibilities: one for magnetic fields applied parallel to the director and one for magnetic fields applied perpendicular to the director (assuming the director is fixed relative to the container). These vary with temperature in liquid crystals in a way similar to the electric susceptibilities, since both depend on how the order

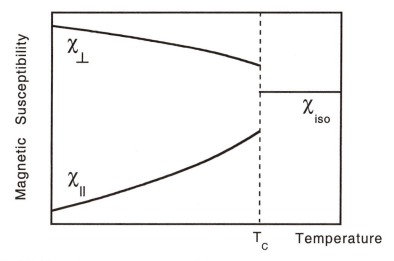

Fig. 3.7. Magnetic susceptibility parallel $(\chi_\parallel)$ and perpendicular $(\chi_\perp)$ to the director for a typical liquid crystal. The dashed vertical line at T_C indicates the transition from nematic liquid crystal to isotropic liquid.

parameter varies with temperature. Data for a typical liquid crystal are shown in figure 3.7.

DEFORMATION OF LIQUID CRYSTALS

Although the director in a liquid crystal frequently is free to point in any direction, there are some important examples where this is not the case. Early scientists were quick to notice that some materials, when placed in contact with a liquid crystal, force the director to point along a specific direction. For example, a glass surface can be prepared in such a way that the liquid crystal molecules near the glass are forced to lie parallel to the surface. The director must therefore lie parallel to the surface, and will remain in this configuration even if an electric or magnetic field is applied. The molecules farther away from the surface are not constrained in this way, so they can respond to electric and magnetic fields. The result of applying a field is to cause the liquid crystal to deform, with the director parallel to the surface in the vicinity of the surface and perpendicular to the surface away from the surface (assuming a field causing alignment of the director perpendicular to the surface). In this situation, the effects of the surface and the electric or magnetic field compete, causing deformation of the liquid crystal. To understand how liquid crystals respond to electric and magnetic fields, therefore, we must also discuss effects related to deformation.

An undeformed liquid crystal is one in which the director points in the same direction throughout the liquid crystal. This is the director configuration that the liquid crystal adopts if it is free from other influences. A deformed liquid crystal is one in which the director changes its direction from point to point. In describing all the ways this might occur, scientists have discovered that all deformations in liquid crystals can be described in terms of three basic types: (1) *splay*, (2) *twist*, and (3) *bend*. These basic deformations are shown in figure 3.8. For all three deformations shown, the director is unchanged in moving into or out of the page. Thus these three basic deformations involve director orientation changes in only two dimensions. In cases of a complicated deformation, more than one of these basic deformations must be used to describe it (for example, both splay and bend in the same plane).

In trying to understand how liquid crystals deform under various conditions, it is useful to think of a deformed liquid crystal as a compressed spring. An outside force is necessary to deform a liquid crystal, just as an outside force is necessary to compress a spring. Once this force is removed, the liquid crystal returns to its undeformed configuration, just as the spring returns to its normal length. The force required to deform a liquid crystal increases as the amount of deformation increases, much as the force required to compress a spring grows as the spring is compressed more and more. When an electric or magnetic field deforms a liquid crystal, the field must supply the force necessary to cause the deformation.

One exception should be pointed out at this time. The molecules that form chiral nematic liquid crystals possess forces between them that on average cause the molecules to orient at a slight angle with respect to each other. This ordering of the molecules causes the director to rotate, giving the liquid crystal a certain amount of twist. For chiral nematic liquid crystals, therefore, the twisted configuration is the spontaneous one; a force is required to

splay   twist   bend

Fig. 3.8. The three basic types of deformation in liquid crystals. Elliptical shapes represent molecules, which become circular when viewed end-on. In all three cases, deformation is present in the plane of the page. There is no change in the director orientation in going into or out of the page.

produce splay or bend or to change the amount of twist from its spontaneous value.

Effects in Thin Samples of Liquid Crystals

In studying liquid crystals, scientists often place a sample between two pieces of glass. This serves to keep the liquid crystal contained in one place, while allowing plenty of light to enter and leave the sample. Although very convenient, this procedure causes the liquid crystal to behave in a complex way, due to the competition between many effects. In this situation, consideration must be given to both the electric or magnetic field effects and the deformation properties in order to gain an understanding of how the liquid crystal behaves.

So delicate is the response of liquid crystal molecules that almost any surface causes the director to orient in a specific direction near the surface. For example, a glass surface need only be rubbed with a piece of cloth before liquid crystal is placed against it in order to force the director to lie along the direction of rubbing. Other effective treatments include the application of a chemical to the glass before adding the liquid crystal. The chemical (called a surfactant) adheres to the glass and produces alignment of the director. Evaporation of solid material onto the glass prior to using it with a liquid crystal also causes alignment of the director. In short, proper preparation of the glass surface can produce alignment of the director in almost any direction. The discussion here is limited to two cases, alignment parallel and perpendicular to the surface of the glass.

Consider what happens when a small amount of nematic liquid crystal is placed between two pieces of glass that have been treated to produce alignment of the director parallel to the surface. Near the two glass surfaces, the director is constrained to point in a certain direction (parallel to the surface). To remain in an undeformed configuration, the liquid crystal between the glass surfaces also adopts a configuration where the director is parallel to the surfaces. This configuration is shown in figure 3.9(a) and is called the *homogeneous texture*. Now imagine that an electric or magnetic field is applied perpendicular to the glass surfaces and that this field tends to orient the director parallel to the field. The molecules near the surfaces are not very free to orient with the field, but the molecules near the middle are fairly free. The electric or magnetic field thus causes the director to change its orientation the most in the middle, with less and less change closer to the surfaces. This deformed structure is shown in figure 3.9(b). The most interesting aspect of this deformation is that it does not occur gradually as the strength of the electric or magnetic field is gradually increased. For fields with strengths below a certain value, the liquid crystal remains in its completely undeformed homogeneous texture. No change at all occurs. Then at some threshold value

Field Off   Field On

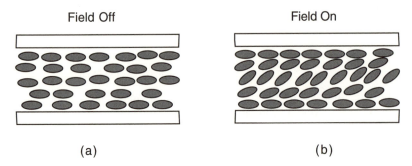

(a)   (b)

Fig. 3.9. Freedericksz transition in the splay and bend geometry. Treatment of the glass orients the liquid crystal molecules parallel to the surfaces, while the electric field tends to align the director perpendicular to the surfaces. The condition when the field is below the threshold value is shown in (a); (b) shows the condition when the field is above the threshold value.

of the field, the deformation begins and then gets greater as the field strength is increased. This transition from an undeformed to deformed texture at a certain value of the field is called the *Freedericksz transition* after the Russian scientist who studied this phenomenon in the 1930s. The Freedericksz transition is not a phase transition, because at any point in the liquid crystal the amount of order of the molecules relative to one another remains the same. It is simply a transition from a uniform director configuration to a deformed director configuration.

From the previous discussion, it is easy to guess what factors must enter into determining the value of the threshold field. Since the anisotropy of the electric polarization or magnetization determines how much tendency there is for the director to orient along the field, the higher the anisotropy of the liquid crystal, the lower the value of the threshold field. Likewise, liquid crystals differ in the magnitude of the force necessary to deform them. Liquid crystals requiring higher forces for deformation obviously have higher threshold field values. Finally, since the deformation occurs continuously across the thickness of the liquid crystal film, the greater the thickness, the less the deformation (change in director orientation per unit distance) at any point. Since the same alignment of the director in the center of the film means less deformation in thicker films, one expects the threshold value for the field to be less in thicker samples. This turns out to be the case. Notice that the first two parameters are properties of the liquid crystal. Once a liquid crystal is chosen for the thin film, these two parameters are set. Only the thickness can then be varied.

The deformation shown in figure 3.9(b) contains both splay and bend. By treating the surfaces of the glass differently and by directing the electric or magnetic field in other directions, it is possible to produce other types of

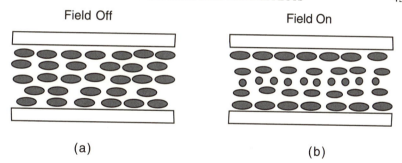

Fig. 3.10. Freedericksz transition in the twist geometry. Treatment of the glass orients the liquid crystal molecules parallel to the surfaces and in the plane of the page, while the electric field tends to align the director perpendicular to the page. The condition with the field below and above the threshold value is shown in (a) and (b) respectively.

deformations. Consider as a second geometry the case shown in figure 3.10. Again the surfaces have been treated for parallel alignment of the director, but now the field is applied into or out of the page. The molecules in the middle of the liquid crystal film tend to point into or out of the page with the field, so the deformed state is one of twist. This is shown in figure 3.10(b). Although the geometry of this situation is different, the changes occur much as in the previous geometry. No deformation takes place at all until the electric or magnetic field strength reaches a certain threshold value, at which time the deformation begins. The value of the threshold field strength depends on the same factors as before, except now the force necessary to cause twist deformation, rather than splay and bend deformation, is important.

A third type of geometry results from treating the glass surfaces in a way that produces perpendicular alignment of the director. Again, with no electric or magnetic field applied, the liquid crystal adopts a completely undeformed director configuration that always points perpendicular to the surfaces. This is shown in figure 3.11(a) and is called the *homeotropic texture*. The electric and magnetic field is applied parallel to the glass surfaces, thus tending to orient the molecules in the middle parallel to the surfaces. The deformed state is shown in figure 3.11(b); notice that the deformation is purely bend. As in the other two geometries, no deformation results as the field is increased until it reaches a threshold value. The value of this threshold field depends on anisotropy and film thickness as before, but is dependent only on the force necessary to produce bend deformation. The transition from undeformed to deformed director configuration is called a Freedericksz transition in all three cases.

One important aspect of the Freedericksz transition is that typical values for the threshold field are not very high. This is due to the relative freedom

Field Off   Field On

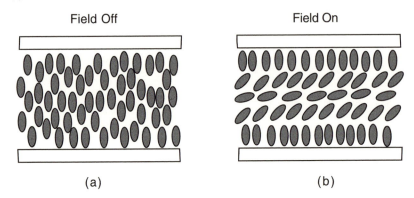

(a)   (b)

Fig. 3.11. Freedericksz transition in the bend geometry. Treatment of the glass orients the liquid crystal molecules perpendicular to the surfaces, while the electric field tends to align the director parallel to the surfaces. The conditions below (a) and above (b) the threshold field strength are both shown.

liquid crystal molecules possess as they diffuse throughout the sample. For a film thickness of 25 micrometers (0.001 inches), the threshold electric field is roughly 400 volts/centimeter, and the threshold magnetic field is about 0.2 teslas. Both of these field strengths are easily produced in the laboratory. In fact, the threshold electric field can be produced by applying a voltage of only 1 volt to electrodes attached to the glass surfaces. The Freedericksz transition is important to the operation of liquid crystal displays, because these transitions cause dramatic changes in the optical characteristics of the liquid crystal film. How light interacts with liquid crystals is one of the most interesting subjects in the field of liquid crystals, and is discussed at length in the next chapter.

HYDRODYNAMIC EFFECTS CAUSED BY ELECTRIC FIELDS

One last example of how liquid crystals delicately respond to electric or magnetic fields is the flow patterns that develop under certain conditions. Motion of the fluid induced by electric fields was noticed by some of the early researchers but was not well understood until the 1960s. The reason for this is that liquid crystal molecules have no net charge, and therefore should experience zero net force in an electric field. An electric field may turn a dipole and orient the liquid crystal molecules, but not cause positional motion of entire molecules in a certain direction. What then causes the motion?

The answer lies in the presence of a small amount of charged impurity ions in the liquid crystal sample. The electric field causes these ions to move from one electrode to the other, but the mobility of these ions is greater parallel to the long axis of the molecules than perpendicular to it. Under

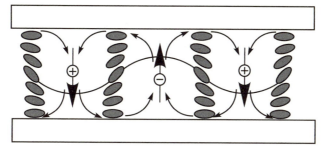

Fig. 3.12. Formation of Williams domains in a liquid crystal cell. Motion of charged impurities (circles and straight arrows) causes the liquid crystal molecules to circulate (curved arrows). This circulation causes a distortion of the director configuration (shown by the wavy line), which focuses light through the sample into alternating bright and dark bands.

some conditions this movement of ions can actually orient the liquid crystal molecules along the electric field, even if an orientation perpendicular to the electric field would otherwise be preferred. In other cases a regular pattern of bright and dark stripes (called *Williams domains*) is produced, and dust particles being carried along by the moving liquid crystal molecules can be seen undergoing periodic motion within the striped bands. This is a more complicated situation, since the interaction between the motion of the ions and the liquid crystal produces a slight separation of the moving positive and negative ions. As shown in figure 3.12, convection cells (shown by the curved arrows) result as the liquid crystal molecules flow in closed patterns in response to this slight separation of the differently charged ions. These convection cells cause a regular distortion of the director throughout the sample, which in turn serves to focus light slightly at regularly spaced points in the liquid crystal. It is this slight focusing of light that produces the pattern of stripes.

If the electric field is strong enough, the flow becomes turbulent. No longer steady and regular, the motion of the molecules at any point in the sample changes erratically. This much more violent motion causes intense scattering of light, and thus the sample appears much brighter and whiter. Scientists refer to this phenomenon as *dynamic scattering*, and it has been the basis for a good number of useful applications.

Chapter 4

LIGHT AND NEMATIC LIQUID CRYSTALS

THE OPTICAL PROPERTIES of liquid crystals are one of the most interesting and certainly the most beautiful features of the phase. How liquid crystals affect light is also the basis for just about all of the applications of liquid crystals. In this chapter we will come to understand what happens to light as it propagates through a sample of nematic liquid crystal. To do this properly, we must first discuss some of the properties of light itself.

LIGHT

Although light has always been a topic of much scrutiny among the civilizations of the world, an understanding of what light actually is became possible only in the late nineteenth century after the development of a new theory for electromagnetism. Our discussion about electric and magnetic fields and how matter behaves in the presence of these fields was based on these same theoretical ideas. Perhaps the most interesting aspect of this theory is the prediction that oscillating electric and magnetic fields can propagate through space without the need for some kind of medium. That is, these oscillating electric and magnetic fields can travel through a perfect vacuum. The scientists of the late nineteenth century saw the connection between this prediction and all the properties of light, so the idea of light as propagating electric and magnetic fields (that is, an *electromagnetic wave*) quickly became a widely accepted notion.

The theory predicts that the electric and magnetic fields in an electromagnetic wave must have certain characteristics. The simplest example is a wave in which the electric and magnetic fields are perpendicular to each other, with both fields perpendicular to the direction the wave is propagating. The strength of both the electric and magnetic fields oscillates as the wave propagates, reaching a maximum value in one direction, then decreasing through zero until it reaches a maximum value in the opposite direction, then decreasing through zero again, and so on. Such an electromagnetic wave is shown in figure 4.1, where a set of axes has also been drawn for reference. This wave has the electric field always directed along the x-axis and the magnetic field always directed along the y-axis, while the wave propagates along the z-axis.

There are two important characteristics of electromagnetic waves that are evident from the figure. First of all, notice that the electric and magnetic

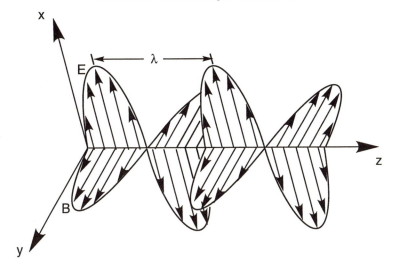

Fig. 4.1. Representation of an electromagnetic wave traveling along the z-axis. The electric field (E) is always parallel to the x-axis, and the magnetic field (B) is always parallel to the y-axis. The wavelength is denoted by the symbol λ.

fields cycle through their values quite regularly. This means that any value of the electric or magnetic field repeats itself after a certain distance along the z-axis. This is shown in figure 4.1 for two places where the electric field reaches a maximum along the positive x-axis. This distance is an important characteristic of any wave and is called its *wavelength*. The distance between two successive crests in an ocean wave is a measure of its wavelength in an analogous fashion. The second important characteristic stems from the fact that the electric field of the wave in figure 4.1 points along a single direction in space. Since the magnetic field must be perpendicular to the electric field, this implies that the magnetic field also points along a single direction in space. This is a special case, so we say that the wave is *linearly polarized*. In stating the direction of *light polarization*, we adopt the convention of choosing the direction of the electric field; the wave in figure 4.1 is linearly polarized along the x-axis.

A word of caution is necessary. In the last chapter we described the "polarization of a sample" as the electric dipole per unit volume. This "polarization" resulted from the application of an electric field to the sample. Here we talk about the "polarization of light" as describing the nature of the electric and magnetic fields of the light. Although scientists use the same word to describe these two phenomena, the two ideas are completely separate. In all cases, the proper meaning of the word "polarization" will be evident from the situation under discussion.

The wavelength of an electromagnetic wave can be as long as hundreds of miles or shorter than the dimensions of a proton. Electromagnetic waves of different wavelengths behave in very different ways, so special names have been given to electromagnetic waves depending on the wavelength. For example, electromagnetic waves with a wavelength greater than 1 meter are called radio waves. Microwaves have wavelengths between 1 millimeter and 1 meter, while X-rays have wavelengths between 0.01 and 1 nanometer (1 nanometer equals 0.000000001 meter). What differentiates light from all other electromagnetic waves is that our eyes are sensitive only to electromagnetic waves within a small range of wavelengths, between 400 and 700 nanometers. Electromagnetic waves with a wavelength in this range constitute light waves, and our eyes can detect them. Within this range of wavelengths, our eyes respond differently to each wavelength, thus causing the sensation of color. An electromagnetic wave with a wavelength of 420 nanometers appears violet to us, while one of 470 nanometers appears blue, 530 nanometers appears green, 580 nanometers appears yellow, 610 nanometers appears orange, and 660 nanometers appears red. We cannot see electromagnetic waves with a wavelength slightly shorter than 400 nanometers or slightly longer than 700 nanometers, so we call them ultraviolet and infrared waves, respectively.

Polarized Light

The linearly polarized wave shown in figure 4.1 is only one example of how electromagnetic waves (light) can be polarized. Another equally simple example would be a wave linearly polarized along the y-axis. In this case the electric field always points along the y-axis, and the magnetic field always points along the x-axis. More complicated examples of polarized light can be understood as a combination of light linearly polarized along the x-axis and light linearly polarized along the y-axis. Before we can appreciate this, however, we must first consider what happens when two electric fields are present simultaneously.

Recall that the concept of the electric field was really a way of describing the electric force. We can therefore create a situation with an electric field along the x-axis and an electric field along the y-axis by situating two charges in the proper locations and asking about the force on a third charge (called a "test" charge). This is shown in figure 4.2(a), where the two arrows attached to the small dot represent the two electric forces on the "test" charge due to the two large charges. The arrows are of different lengths because the forces due to each of the larger charges are not equal. The total force on the smaller charge is just the sum of both forces, making sure we take the direction of the two forces into account. A convenient way of doing this is to construct the rectangle defined by the two forces, since the total force is just along the

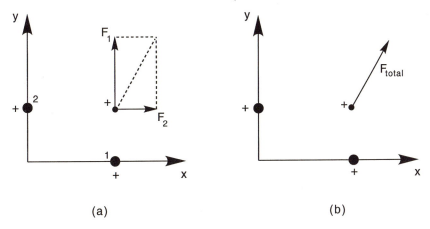

Fig. 4.2. The electric force on a "test" charge (small dot) due to two other charges (large dots). F_1 is the force due to charge 1 and F_2 is the force due to charge 2. The total force (sum of F_1 and F_2) is shown in (b).

diagonal. This procedure is shown in figure 4.2(b). If we wish to describe the electric field at the location of the smaller charge, therefore, we would simply say it points along the direction of the total force in figure 4.2(b), since by definition the direction of the electric field is the same as the direction of the electric force.

We can also arrive at the same result through a consideration of the electric fields due to the two charges rather than the force they produce. Figure 4.3(a) shows the electric fields due to the two larger charges at the point in space where the "test" charge is located. The different lengths of the two arrows again indicate that the strengths of the two fields are not equal. Notice that forming the rectangle using these two fields, as in figure 4.3(b), produces the same electric field direction as before. Therefore, when two electric fields are present along the x-axis and y-axis, the total electric field at any point can be obtained by drawing the diagonal for the rectangle formed by the two electric fields.

The procedure just described is valid for adding together any two quantities that possess directions at right angles to each other. For example, imagine you are in a large city and go for a walk. First you walk four blocks east. On a map of the city, this walk can be represented by an arrow to the right four blocks long. Next you walk five blocks north. This is represented by an arrow five blocks long pointing toward the top of the map. The addition of these two "walks" can be represented by an arrow from your starting position on the map to your ending position. This is simply the diagonal of the rectangle formed by the east- and north-pointing arrows. You can prove to yourself that the "walk" represented by this diagonal actually equals the sum of the

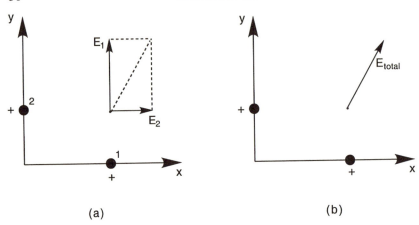

Fig. 4.3. The electric field at a point in space due to two charges. E_1 is the electric field due to charge 1 and E_2 is the electric field due to charge 2. The total electric field (sum of E_1 and E_2) is shown in (b).

two other "walks," because you know that if you could take this diagonal "walk" (the buildings do present somewhat of a hazard), you would end up at the same position as after taking the east and north "walks." In analogous fashion, the diagonal electric field is identical to the sum of the two fields along the x-axis and y-axis.

This technique can be used to determine what happens when linearly polarized light along both the x and y directions is present. Let us first examine the case when the maximum strengths of the two electric fields are equal. That is, the electric fields of both waves oscillate, but the maximum value of the electric field along the x-axis and the maximum value of the electric field along the y-axis are equal. Now imagine sitting at a point along the z-axis as these two electromagnetic waves pass. The electric fields of both waves will oscillate as the waves pass, starting from zero, reaching a maximum in one direction, going through zero again, reaching a maximum in the other direction, and so on. To understand this situation, we must keep track of what the electric fields along both the x-axis and y-axis are doing. Figure 4.4(a) shows the strength of both electric fields at eight different times as the two waves pass. The numbers on each of these individual electric fields give the chronological order. That is, the electric fields of both waves start at zero strength (1), reach a maximum in one direction (3), pass through zero again (5), reach a maximum in the other direction (7), and finally reach zero again. The electric fields at times intermediate to these times (2, 4, 6, 8) are also shown. Let us add together the two electric fields at each of these eight points in time. This is shown in figure 4.4(b), where the different rectangles have been drawn. The numbers again give the chronological order.

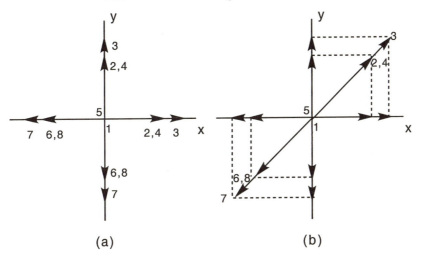

(a)   (b)

Fig. 4.4. The combination of two electromagnetic waves with polarizations along the x-axis and y-axis. The electric field strengths for each of the waves are given in (a). The numbers represent different times in chronological order. The combined electric field strengths are given in (b) for the same eight times. The phase difference between the two waves is 0°.

Notice that the electric field simply oscillates along a line at 45° to the x-axis and y-axis. The combination of these two linear polarizations is just linearly polarized light along a different direction.

In the previous example, the electric fields of the linearly polarized waves both started out at zero and increased in the positive x and y directions. Consider what occurs if they start out at zero again, but with the light linearly polarized along the y-axis now increasing in the opposite direction. Figure 4.5(a) shows the individual electric fields at the same eight times, and figure 4.5(b) gives the result. The numbers again denote the chronological order as the two waves pass. The result is again linearly polarized light, but now polarized along a different direction.

Consider one more case. What happens if the electric field polarized along the x-axis reaches its maximum value along the positive x-axis just as the electric field polarized along the y-axis starts to increase from zero along the positive y-axis? We can apply the same procedure as before, but the result is very different. As can be seen from figure 4.6, the total electric field does not oscillate along a single direction, but rotates counterclockwise, always maintaining the same strength. We call this *circularly polarized* light, as opposed to linearly polarized light. To indicate that circularly polarized light can rotate in either of two directions, we call light with an electric field rotating counterclockwise at some point in space (as seen by an observer

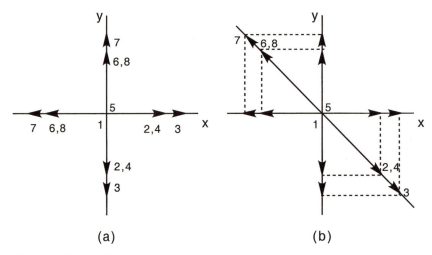

Fig. 4.5. The combination of two electromagnetic waves with polarizations along the x-axis and y-axis. The electric field strengths for each of the waves are given in (a). The numbers represent different times in chronological order. The combined electric field strengths are given in (b) for the same eight times. The phase difference between the two waves is 180°.

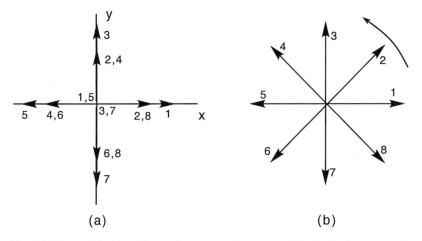

Fig. 4.6. The combination of two electromagnetic waves with polarizations along the x-axis and y-axis. The electric field strengths for each of the waves are given in (a). The numbers represent different times in chronological order. The combined electric field strengths are given in (b) for the same eight times; the result is circular polarization rotating in the direction of the curved arrow. The phase difference between the two waves is 90°.

facing the source of light) left circularly polarized light, and light with an
electric field rotating clockwise (as seen by a similar observer) right circularly
polarized light. To produce right circularly polarized light, the electric field
polarized along the y-axis must reach its maximum value along the positive
y-axis just as the electric field polarized along the x-axis starts to increase
from zero along the positive x-axis.

These last three examples differ only in what each of the two linearly
polarized waves is doing relative to the other. We use the concept of *phase
difference* to describe these differences. For example, in the case shown in
figure 4.4 both polarized waves are doing the same thing at the same time. We
say there is zero phase difference between these two waves or, alternatively,
we say the two waves are in phase. In the example of figure 4.5, the electric
field polarized along the y-axis is half a cycle ahead of the electric field
polarized along the x-axis. Since a full cycle can be represented by 360°,
we say there is a 180° phase difference between the two waves, or, since the
two waves are behaving in opposite fashion, we can also say the two waves
are out of phase. The example of left circular polarized light (figure 4.6)
by the same reasoning represents two waves 90° out of phase. Whether the
two waves combine to give right or left circularly polarized light depends on
which electric field is a quarter cycle ahead of the other.

In all these examples, the maximum strengths of the two linearly polarized
electric fields are equal. When this is not the case, (1) two waves in or
out of phase give linearly polarized light at some angle other than 45° to
the x-axis and y-axis, and (2) two waves 90° out of phase give elliptically
polarized light (the electric field strength varies as it rotates). In the latter
case, the ellipse outlined by the electric field as it rotates has its long axis
along the x-axis or y-axis. A phase difference of other than 0°, 90°, or 180°

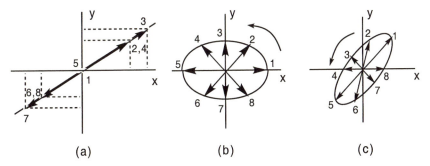

(a)   (b)   (c)

Fig. 4.7. The combination of two electromagnetic waves with polarizations along the
x-axis and y-axis as in figures 4.4–4.6, but with unequal maximum electric field
strengths. The phase difference between the two waves is 0° in (a), 90° in (b), and
between 0° and 90° in (c).

also produces elliptically polarized light, but now the long axis of the ellipse will not be along either the x-axis or y-axis. All three of these cases are shown in figure 4.7.

HOW LIGHT INTERACTS WITH MATTER

Because my discussion of light as an electromagnetic wave has assumed there is nothing around to affect the wave, the results really apply to light propagating through a vacuum. In this case, light travels at a speed of 300,000,000 meters per second (186,000 miles per second). The presence of matter, however, affects the electromagnetic wave in several important ways. Just as a steady electric or magnetic field causes matter to become polarized or magnetized, the oscillating electric and magnetic field of a light wave produces oscillating motion of the charges in a material. These oscillating charges in turn produce their own oscillating electric and magnetic fields, resulting in a combined electromagnetic wave different from the one that would have been present if no matter were around. Although this situation is in principle quite complicated, the additional electric and magnetic fields coming from all the different charges in the material cancel themselves out, except for those fields that are propagating in the same direction as the original wave. These induced electromagnetic waves combine with the original wave to produce an electromagnetic wave that travels through the material with a velocity slower than the velocity of light in a vacuum. How much slower light propagates in a material depends on the exact nature of the charges in the material. We therefore assign a number to each material that represents the factor by which the velocity of light is slowed down. That is, if v is the velocity of light in a material and c is the velocity of light in a vacuum, then this factor, n (called the *index of refraction* of the material), is defined as

$$n = c/v.$$

It is clear from this definition that the index of refraction of all materials must be greater than one. For example, the index of refraction is 1.0003 for air, 1.33 for water, and about 1.5 for glass.

Because light of various wavelengths (as measured in a vacuum before it enters matter) produces different effects on the charges present in matter, the velocity at which light propagates through a material depends on the wavelength of the incident light. This means that the index of refraction also depends on wavelength. Usually this effect is small; for example, glass might have an index of refraction of 1.51 for red light and 1.53 for violet light. In some cases, however, this small effect can lead to dramatic results.

Light traveling through a material also tends to lose intensity as the material absorbs some of the energy of the electromagnetic wave. The ability of a material to absorb energy from the light wave also depends on the wavelength of the incident light, but in this case the effect can be quite large. For example, a piece of red glass in a stained glass window absorbs much more light of shorter wavelength (blue and green) than of longer wavelength (red). Although light of many wavelengths (white light) strikes the glass, much more of the red light gets through, so we see a red color. The absorption properties of materials vary considerably, resulting in a wealth of possible effects.

The interaction of light with matter is even more interesting at the point where light goes from one material to another. Because the arrangement of charges in the two materials is different, the oscillating electric and magnetic fields from these charges are different. Although these fields cancel themselves except in the forward direction within each material, they do not cancel in the same way at the boundary between the two materials, because each material produces different oscillating electric and magnetic fields. The oscillating electric and magnetic fields produced at the boundary between two materials together with the original electromagnetic wave now cancel except for those traveling in two directions. One of these directions is backward at an angle, thus producing a *reflected* wave. The second direction is forward, but in a direction different from the direction of the wave in the first material. This is illustrated in figure 4.8. The bending of light as it passes from one material to another is called *refraction*. For a given angle between the boundary and the direction of propagation of the wave in the first material, the new angle in the second material depends only on the indices of refraction of the two materials. The amount of light reflected as opposed to refracted at the boundary, however, also depends on how the light is polarized.

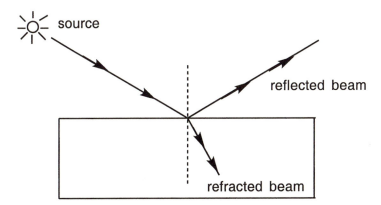

Fig. 4.8. Reflection and refraction of light at the surface of an isotropic material. The paths of the reflected and refracted light are independent of the polarization of the light.

BIREFRINGENCE OR DOUBLE REFRACTION

In the preceding discussion, we assigned a single index of refraction to each material. A simple experiment quickly indicates that this is not always valid. Let us take a beam of light that is linearly polarized along the x-axis and direct it at an object, measuring the angle the beam makes on either side of the boundary between the air and object. Measurement of these angles allows us to find the index of refraction for the object. Just to be sure everything is working well, let us repeat the measurement for each object using light linearly polarized along the y-axis. For many objects (glass, water, a large salt crystal), both measurements give the same number for the index of refraction, but other objects (a large crystal of quartz or calcite) produce two different numbers. This means that light propagating through these latter materials is traveling at a different velocity depending on whether it is polarized along the x-axis or y-axis. As previously discussed in chapter 3, a material that has different properties for different directions is called an anisotropic material. Obviously, the first set of materials must be isotropic and possess one index of refraction, while the second set must be anisotropic and possess more than one index of refraction. This effect is easily seen when a beam of light containing both x-polarized and y-polarized light is incident on an anisotropic material. The two polarizations have different indices of refraction and therefore travel along two different directions inside the material. This phenomenon is called *birefringence* or *double refraction* and is illustrated in figure 4.9.

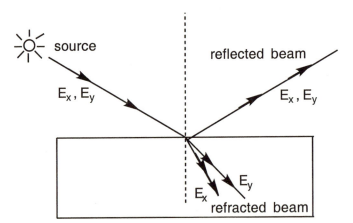

Fig. 4.9. Reflection and refraction of light at the surface of an anisotropic material. The path of the reflected light is independent of the polarization (E_x or E_y) of the light. The path of the refracted light depends on whether the light is linearly polarized along the y-axis (out of the page) or x-axis (perpendicular to both the y-axis and the direction of light propagation).

All effects that depend on the index of refraction are different for these two polarizations. For example, the amount of light reflected or transmitted at the interface depends on the polarization of the light. Likewise, the amount of light absorbed by the material is different for the two polarizations. In fact, it is possible to fabricate materials that absorb one polarization nearly completely while allowing most of the other polarization to pass through. Such materials can be used to produce linearly polarized light, since only one polarization emerges when they are illuminated by unpolarized light. Similarly, these "polarizers" can also be used to check if a beam of light is polarized, in that one orientation of the "analyzer" blocks all the light if the beam is linearly polarized. The production and analysis of polarized light using such materials is shown in figure 4.10.

Birefringent materials possess one additional important property. Consider light containing both x-polarized and y-polarized components that strikes a slab of such a material at an angle of 90° to the surface. In this case neither polarization is bent, so both polarized components travel along the same direction in the material, but at different velocities. One of the polarizations therefore gets ahead of the other, causing the phase difference between the two polarizations to change as the light travels through the material. If the two polarizations started out in phase (zero phase difference), they will emerge with a phase difference that depends on the thickness of the material. Devices that utilize this effect are called *phase retarders* and are extremely useful in many optical applications. For example, if the thickness is carefully adjusted to cause a 90° change in the phase difference, light that is linearly polarized upon entering the material emerges as circularly polarized light. Likewise, circularly polarized light incident on this slab emerges as linearly polarized light. Obviously, varying the thickness of the material produces many other possibilities.

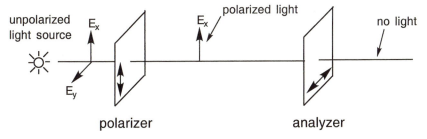

Fig. 4.10. The effect of a crossed polarizer and analyzer. The polarizer allows only light polarized along the x-axis to pass, while the analyzer allows only light polarized along the y-axis to pass. Since there is no light polarized along the y-axis emerging from the polarizer, no light passes through the analyzer.

BIREFRINGENCE IN NEMATIC LIQUID CRYSTALS

The anisotropy of nematic liquid crystals causes light polarized along the director to propagate at a different velocity than light polarized perpendicular to the director. Nematic liquid crystals are therefore birefringent. This fact is never more obvious than when a liquid crystal is placed between crossed polarizers. Normally no light emerges from crossed polarizers, because the light emerging from the first polarizer is completely absorbed by the second polarizer. Insertion of an isotropic material does not change this because the polarization of the light is unchanged as it travels through an isotropic material. Now consider the case where the polarized light from the first polarizer is oriented such that the direction of polarization makes an angle other than 0° or 90° with the director of the liquid crystal. From our past discussion, we know we can consider this light to be composed of light polarized along the director and light polarized perpendicular to the director with zero phase difference (figure 4.4). In passing through the liquid crystal, the two polarizations get out of phase and in general emerge as elliptically polarized light. Since the electric field of elliptically polarized light is constantly rotating completely around during each cycle, it is parallel to the polarization axis of the second polarizer twice during each cycle. Some light will therefore emerge from the second polarizer. Introduction of a liquid crystal between crossed polarizers in general causes the field of view to appear bright, whereas with no liquid crystal between the polarizers it is dark. It is this effect that was discovered by Otto Lehmann and led him to the correct conclusion that liquid crystals must be anisotropic.

Although liquid crystals normally appear bright when viewed between crossed polarizers, there are two conditions under which they continue to appear black. If the polarized light incident on the liquid crystal has its polarization direction either parallel or perpendicular to the director, all the light is polarized along this one direction in the liquid crystal, and there is no need to consider light polarized at 90° to this direction. Since only one polarization is present, it propagates through the liquid crystal at one velocity and emerges polarized along the same direction. It is therefore extinguished by the second polarizer.

Pictures taken of liquid crystals under the microscope are usually taken with the sample between crossed polarizers. The director usually points in different directions at different points within the sample. Areas where the director is oriented parallel or perpendicular to the axes of the polarizers appear dark, while areas where the director makes an angle with the polarizer axes other than 0° or 90° appear bright. This is best illustrated in plate 1. Examination of this plate and others reveals an additional fact: there are many places where the brightness changes abruptly, indicating that the orientation of the director must also change abruptly there. These lines are called

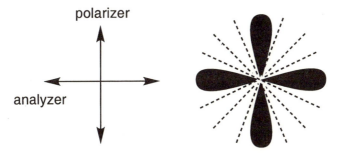

Fig. 4.11. Bright and dark areas around a defect in a liquid crystal viewed with the sample between a crossed polarizer and analyzer. The director (dashed lines) always points away from the defect point. No light emerges from the analyzer when the director is parallel to either the polarizer or analyzer axis.

disclinations and represent places where the director is really undefined, since it points in many directions within an extremely small region. These disclinations are therefore defects and are interesting in their own right, as will be seen in a later chapter.

Plate 1 reveals that point disclinations also exist. At many places in the picture four dark areas converge at right angles. Since dark areas occur when the director is along one of the polarizer axes, these four dark areas must define the two axes of the crossed polarizers. One director configuration that has four areas where the director is parallel to the axes of the crossed polarizers is where the director always points away from a point. This is shown schematically in figure 4.11. Although the director points away from the center of each of the crosses in plate 1, the orientation of the dark areas is determined by the crossed polarizers. That is why the dark areas at the point of convergence are all oriented in the same direction. These convergence points in the picture represent defects embedded in regions where the director is otherwise well defined.

Because the degree of orientational order in a liquid crystal varies with temperature, the indices of refraction for light polarized parallel and perpendicular to the director also change with temperature. At higher temperatures, where the order parameter is smaller, the two indices are closer together in value. The opposite is true at lower temperatures in the liquid crystal phase, where the order parameter is larger. The data for a typical liquid crystal are shown in figure 4.12. Notice that at the transition to the isotropic liquid the anisotropy disappears, and one index of refraction serves for all polarizations.

THE SCATTERING OF LIGHT BY LIQUID CRYSTALS

One of the most obvious characteristics of liquid crystals is that they appear cloudy in bulk quantities, whereas isotropic liquids appear clear. The reason

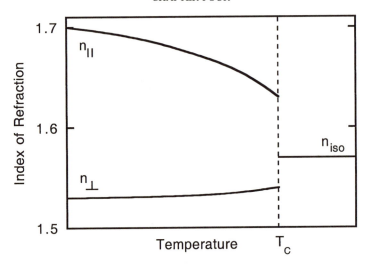

Fig. 4.12. Indices of refraction for light polarized parallel $(n_\parallel)$ and perpendicular $(n_\perp)$ to the director of a typical liquid crystal. The dashed vertical line at T_C denotes the phase transition from liquid crystal to isotropic liquid.

for this lies again with the fact that light is an electromagnetic wave and therefore interacts with matter as it propagates.

We saw before that light propagates only in the forward direction as it travels inside a material. Only at boundaries between two different materials can its path be changed. This is true provided there are no changes going on in the electric or magnetic properties of the material. For example, if the electric susceptibility at some point in a substance is changing in time as an electromagnetic wave passes, the polarization produced by the wave is not as simple as before. This more complicated response plus the original wave can cause a new electromagnetic wave to propagate in a direction other than the forward direction. This produces a *scattered* wave and is possible only as long as these changes are taking place in the substance.

To illustrate this mechanism, consider the case of an air gun shooting a stream of Ping-Pong balls at a stationary Ping-Pong paddle. After striking the paddle, each Ping-Pong ball is reflected in the same direction, producing a stream of balls representing the reflected wave. Obviously, in this case there is no transmitted (refracted) wave. Notice, however, that only one reflected wave is possible. Now imagine moving the paddle randomly as the stream of balls strikes the paddle. This causes the velocity of the paddle to change with time, and you can imagine that the result is streams of Ping-Pong balls reflected in many directions. These various streams represent scattered waves, which exist only as long as the orientation of the paddle varies in time.

Because the director of a liquid crystal responds so sensitively to changes of any kind, its orientation tends to fluctuate slightly simply due to the nearly random motion of the molecules. This causes the orientation of electric polarization to fluctuate, which scatters light incident on the liquid crystal. It is this scattered light that makes the liquid crystal appear cloudy. Isotropic liquids do not scatter light, because the lack of a preferred direction for the molecules means there is no anisotropy in the electric polarization and therefore nothing that can fluctuate.

Smectic liquid crystals possess an additional mechanism by which they scatter light. The layer structure of smectics means that the electric and magnetic properties of the liquid crystal also depend on position, being different within a smectic layer than between two layers. The same nearly random motion of the molecules causes the layers to both undulate and vary in thickness slightly in a fluctuating fashion. In smectic liquid crystals, therefore, light is scattered by fluctuations in both the director and the layer structures.

Chapter 5

LIGHT AND X-RAYS IN OTHER TYPES OF
LIQUID CRYSTALS

THE TWISTED STRUCTURE of chiral nematic liquid crystals and the layered structure of smectic liquid crystals affect the propagation of light in ways quite different from what occurs in nematic liquid crystals. We shall see in this chapter that strikingly beautiful effects occur in chiral nematic liquid crystals when the pitch of the helical structure is about equal to the wavelength of light in the liquid crystal. This phenomenon forms the basis for a number of useful applications. Likewise, similar effects occur in smectic liquid crystals for electromagnetic waves with a wavelength about equal to the layer thickness. This thickness is significantly smaller than the wavelength of visible light but falls within the X-ray part of the electromagnetic spectrum.

CIRCULAR BIREFRINGENCE

As discussed in chapter 1, the director in a chiral nematic liquid crystal is not constant but rotates about an axis in a helical fashion. This additional structure interacts with light propagating through the liquid crystal in a way very different from the structure in nematics, producing some of the most interesting optical effects in liquid crystals.

Let us assume the director rotates about the z-axis. This means that at some points in the liquid crystal the director points along the x-axis, and at other points it is aligned with the y-axis. In fact, it rotates very uniformly through all directions in the xy plane. This situation is illustrated in figure 5.1, where the director is shown as an arrow pointing in different directions from different points along the z-axis. If you follow the head of this arrow as it moves along the z-axis, it outlines a helix that is shown in the figure as a curved line. This helix is easily visualized it you imagine being a flea walking along the threads of a screw. If you walk in the same direction along a single thread, your path will be the same as the helix drawn in figure 5.1. Perhaps the most important property of any helix is the distance along the z-axis for one full turn of the helix. As mentioned in chapter 1, this is called the pitch of the chiral nematic liquid crystal and is labeled in the figure as P. In our analogy of the flea walking along a single thread of a screw, the pitch would

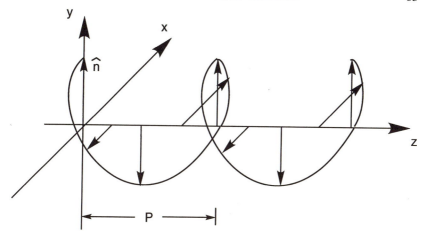

Fig. 5.1. Director configuration in a chiral nematic liquid crystal. The arrows along the z-axis represent the director, which is oriented differently for regions of liquid crystal along the z-axis. P is the pitch of the structure, and the curved line is the helical path of the tip of the arrow. The helix is right-handed (see text).

be the distance from one thread to the next measured parallel to the long axis of the screw.

Consider light linearly polarized along the x-axis propagating along the z-axis of a chiral nematic liquid crystal. This light must interact with parts of the liquid crystal having a director oriented at all angles with respect to the polarization of the light. Now consider what happens to light linearly polarized along the y-axis. It too must interact with parts of the liquid crystal having a director oriented at all angles relative to the polarization axis. In fact, light linearly polarized along any direction in the xy plane must interact with parts of the liquid crystal having a director oriented at all angles relative to the polarization axis. This means that linearly polarized light propagates along the z-axis with the same velocity regardless of its axis of polarization. Since components of light polarized along both the x-axis and y-axis travel together, no change in the phase difference between these two components results. This is different from what happens in nematic liquid crystals, where a change in the phase difference between the components polarized along the x-axis and y-axis normally occurs.

The situation is quite different for circularly polarized light. Just as circularly polarized light can be either left-handed or right-handed, the structure of a chiral nematic liquid crystal can be either right-handed or left-handed. The definition, however, is slightly different from the definition used for circularly polarized light. Imagine looking back along the z-axis, observing the director for parts of the liquid crystal that are closer and closer to you. The director appears to rotate either counterclockwise or clockwise. If it seems to rotate counterclockwise, we call it a right-handed helix; if it appears to rotate clockwise, it is a left-handed helix. Returning to our analogy of the thread of a screw, imagine looking along the axis of a screw as a nut is rotated so that it approaches your eye. Is it rotating clockwise or counterclockwise? For most screws, the nut rotates counterclockwise, indicating that most screws are right-handed. Notice that it makes no difference which end of a screw you use to start the nut. A right-handed screw is simply right-handed, meaning that the nut must rotate counterclockwise if it is to move closer to your eye.

Now consider what happens when circularly polarized light passes through a chiral nematic liquid crystal. There are two possibilities: either the light and the liquid crystal have the same sense of chirality (right-handed or left-handed), or they have the opposite sense of chirality. The interactions of the light with the chiral structures are different in these two cases, and this gives rise to two different propagation velocities. It is like trying to twist a nut on to a screw. Once again there are two situations. Either the nut and screw have the same chirality or opposite chirality, and the success you have in getting the nut to twist onto the screw is very different for the two cases. Because of the two different velocities, we can assign an index of refraction for right circularly polarized light, n_R, and another index of refraction for left circularly polarized light, n_L. The values of these two indices are equal for nematic liquid crystals, but unequal for chiral nematic liquid crystals. We could describe this situation by saying that chiral nematic liquid crystals possess *circular birefringence* in much the same way we said nematic liquid crystals possess linear birefringence.

In general, this circular birefringence depends on the amount of orientational order, so it too is greater at lower temperatures and less at higher temperatures. Even more important, however, is the fact that the difference between n_R and n_L varies with wavelength in a much more dramatic fashion than the linear birefringence of nematics does. The reason for this will be clear later.

OPTICAL ACTIVITY

In analyzing what happens to light as it propagates through a nematic liquid crystal, we considered the incoming light to be made up of lightwaves polarized along the x-axis and others polarized along the y-axis, and then followed what happened to each kind of polarization. By adding the two together after they propagated through the liquid crystal, we were able to determine the polarization of the light as it emerged. For chiral nematic liquid crystals, we must do the same thing, but instead of thinking about the incoming light as made up of two linear polarizations, we must imagine the incoming light as composed of the two possible circular polarizations, right and left.

Let us consider the simple case of equal amounts of right and left circularly polarized light traveling down the z-axis. For the moment, assume no liquid crystal is present, so that the two polarizations travel with the same velocity. A view of this situation is given in figure 5.2, where the view is for an observer situated at some point on the z-axis looking backward along the z-axis as the light passes. From this vantage point at a fixed location, the electric field rotates clockwise for the right circularly polarized light and counterclockwise for the left circularly polarized light as time passes. This is shown in figure 5.2 by labeling the positions of the electric field for the two polarizations with numbers indicating the sequence in time. Notice that at every time in the sequence the two electric fields are like mirror images of each other, directed equally on either side of the y-axis. There is therefore no net electric field along the x-axis, so the total electric field must always point along the y-axis. The light must be linearly polarized along the y-axis. Equal amounts of right and left circularly polarized light therefore produce linearly polarized light in a medium with no circular birefringence.

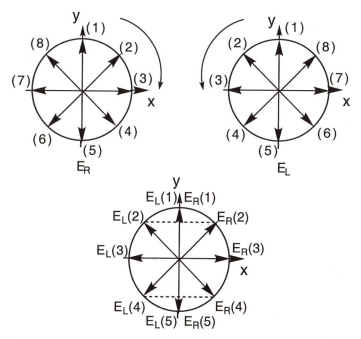

Fig. 5.2. Addition of right- (E_R) and left-handed (E_L) circularly polarized light. The upper two diagrams show the electric field direction at eight times for each of the circular polarizations. The numbers give the times in chronological order. The lower diagram shows how these polarizations combine. They cancel each other along the x-axis and produce an electric field that always points along the y-axis.

What happens as this light that is linearly polarized along the y-axis enters a chiral nematic liquid crystal? The electric fields for the right and left circular polarizations continue to rotate, but because they travel with different velocities, one gets ahead of the other. Let us assume the left circularly polarized light travels slower than the right, so that after traveling through a chiral nematic liquid crystal of thickness d, the right circularly polarized light has gotten ahead of the left circularly polarized light. What emerges at any time, therefore, is left circularly polarized light combined with right circularly polarized light that entered the liquid crystal after it. Thus the right circularly polarized light is at a more advanced angle than the left circularly polarized light. This is shown in figure 5.3 by assuming the left circular polarization emerges along the y-axis, with the right circular polarization ahead of the y-axis by an angle θ. As the light continues to pass by this point, the electric fields for the two polarizations continue to rotate, but now in exactly opposite fashion. As can be seen in figure 5.3, the two electric

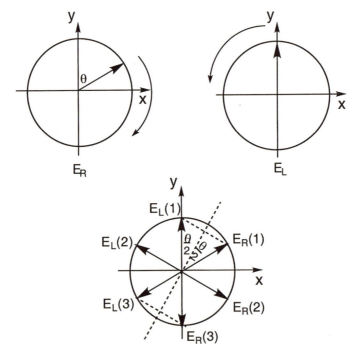

Fig. 5.3. Addition of right- and left-handed circularly polarized light when the electric field for the right-handed polarization is ahead of the left-handed polarization by an angle θ. The upper two diagrams show the two polarizations while the lower diagram reveals how they combine to produce an electric field that always points along a line that makes an angle of $\theta/2$ with the y-axis.

fields now fall at every time in the sequence on opposite sides of an axis inclined relative to the y-axis. The light must therefore be linearly polarized along this direction. The chiral nematic liquid crystal has caused the axis of linear polarization to rotate clockwise as seen by an observer looking at the light as it approaches. Since the angle through which the polarization is rotated depends on how far behind the left polarization gets relative to the right polarization, this angle must be proportional to the thickness of the chiral nematic liquid crystal. The value of the angle divided by the thickness of the sample therefore does not depend on the thickness of the sample. We define this ratio as the *optical activity* of the chiral nematic liquid crystal. Chiral nematics typically have quite a high amount of optical activity, with values of 300°/mm quite standard.

If instead of the situation just described, we imagine the case where the right circular polarization travels at a slower velocity than the left circular polarization, it is not difficult to see that the linear polarization axis will be rotated in a counterclockwise fashion for an observer watching the light emerge from the liquid crystal. Counterclockwise rotation of the axis of polarization is assigned a negative value of optical activity, while clockwise rotation of the axis of polarization is assigned as positive value. Positive values for the optical activity result when the left circular polarization travels slower than the right, while the opposite situation leads to negative values for the optical activity.

Constructive Interference

Previously I discussed how electromagnetic waves are emitted by the charges in a material as an electromagnetic wave passes through it. Because all parts of the material are emitting waves, they tend to cancel except for waves traveling in the same direction as the original wave. This is true unless the electric or magnetic properties of the material themselves vary from one place to another, since in such cases the waves being emitted from different parts of the sample are not all the same. A particularly fascinating case occurs when the electric or magnetic properties of the material repeat themselves over a distance equal to half the wavelength of the light. As shown in figure 5.4, this leads to the situation where the waves emitted by identical regions of the material add together in the backward direction.

The fact that the identical regions are separated by a distance equal to half the wavelength (denoted by $\lambda/2$ in figure 5.4) causes the electric field of each wave traveling backward to be in phase as it emerges from the sample. The waves therefore add together instead of canceling. This phenomenon is called *constructive interference*, and the result is that a sizable electromagnetic wave is not only transmitted but also reflected.

Anyone who has watched a marching band perform various routines can understand why constructive interference occurs. Imagine a band marching

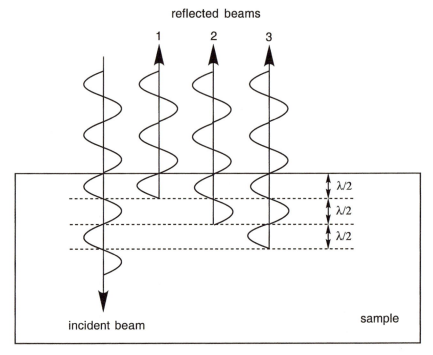

Fig. 5.4. Constructive interference when identical regions in the sample are separated by half a wavelength of light, $\lambda/2$. All the reflected light is in phase, thus generating a strong reflected beam.

along in rows of people separated by four paces. Because the formation repeats itself over a distance of four paces, we can make the analogy that the band represents a wave with a wavelength equal to four paces. At some point the first person in the first column reverses his or her direction (stepping to the side slightly to avoid a collision), and the following people in the first column do the same thing at the same place. If the people in the second column perform the same maneuver at a point two paces farther along than the first column, the two columns will still be arranged in rows as they march in the opposite direction. The other columns can perform the maneuver at points separated by two paces, with the result that the entire band will end up marching in the opposite direction but still arranged in rows. The band members in each row have changed, but arrangement of members into rows with four paces in between is preserved. Just as in the case of electromagnetic waves, the band columns started out in phase with each other and ended up in phase with each other. This would not have been the case if the columns had performed their maneuvers at points separated by one or three paces.

SELECTIVE REFLECTION

Constructive interference occurs in chiral nematic liquid crystals when the pitch of the helical structure is equal to the wavelength of light in the liquid crystal, since the helical structure repeats itself over a distance equal to half the pitch. If light with various wavelengths (white light, for example) is incident on a chiral nematic liquid crystal, most of it will be transmitted with some optical activity—except for light with a wavelength in the liquid crystal equal to the pitch. This phenomenon is called *selective reflection*, because only one wavelength is reflected. If this wavelength falls in the visible range, the light will possess a certain color. For this reason chiral nematic liquid crystals often appear brightly colored in reflection, with the color determined solely by the pitch of the liquid crystal.

Imagine an experiment in which equal amounts of right and left circularly polarized light are incident on a chiral nematic liquid crystal. If the colored light being reflected is examined, it is found that it is either right or left circularly polarized, depending on whether the helix of the chiral nematic is right- or left-handed. The fact that the repeating structure is itself right- or left-handed produces the additional effect that the constructive interference works for one circular polarization but not the other. Figure 5.5 illustrates the results of an experiment similar to the one imagined. Notice that a very narrow range of wavelengths is reflected and only for one polarization of light.

One interesting example of selective reflection in nature is the color of some beetles. At some stage in the development of the cuticle of these beetles, a liquid crystalline substance is secreted. This substance then hardens, with the molecules fixed in place with the orientational order of the chiral nematic phase. The pitch of the helix determines what color of light is

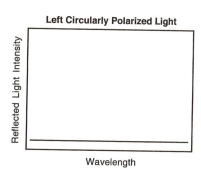

Fig. 5.5. Typical results of an experiment that measures the light reflected from a chiral nematic liquid crystal. Only one circular polarization is reflected and only when the wavelength of the light in the liquid crystal equals the pitch.

selectively reflected, and as expected, the light reflected from these beetles is circularly polarized.

ANOMALOUS OPTICAL ACTIVITY

The reflection of one circular polarization also has a dramatic effect on the optical activity. Of course, for thick enough samples and the proper wavelength of light, only one circular polarization emerges from the liquid crystal since the other is completely reflected. In this case the optical activity cannot be measured. For wavelengths near this value, however, the index of refraction for the polarization being reflected changes abruptly from a value much smaller than the index for the opposite polarization to a value much larger than the other index. Since the sign of the optical activity depends on which index is larger, this means that the optical activity abruptly changes from one sign to another at this wavelength. In addition, the size of the optical activity depends on the difference between the two indices of refraction, so this change in sign is actually from a very large value of the optical activity of one sign to a very large value of the other sign. Data typical of a liquid crystal are plotted in figure 5.6, clearly showing the dramatic effect for light of a particular wavelength. This anomalous behavior can produce extremely large rotations of the axis of linear polarization in very thin samples. For example, rotations of 10,000°/mm are quite common.

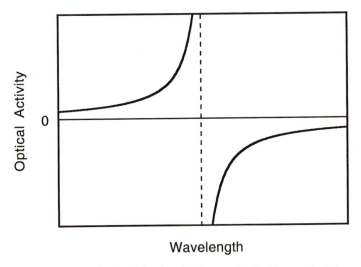

Fig. 5.6. Anomalous optical activity in chiral nematic liquid crystals. The optical activity changes abruptly when the wavelength of the light in the liquid crystals equals the pitch.

CHIRAL NEMATICS BETWEEN CROSSED POLARIZERS

Nematic liquid crystals appear bright between crossed polarizers due to the fact that the linear birefringence changes the linearly polarized light to elliptically polarized light, some of which is transmitted by the second polarizer. Since the amount of linear birefringence is somewhat wavelength dependent, more light at one end of the spectrum is transmitted relative to light at the other end of the spectrum. It is therefore possible for nematic liquid crystals to appear slightly colored when viewed between crossed polarizers.

The situation is quite different for chiral nematic liquid crystals. The first polarizer ensures that linear polarized light enters the sample. The optical activity of the chiral nematic, however, rotates the axis of polarization so that some light can get through the second polarizer. The more the rotation of the axis of polarization, the brighter the sample appears. But we have just seen that wavelengths of light in the sample that nearly equal the pitch are rotated the most. This light will be transmitted through the second polarizer much more than other wavelengths, producing a specific color when viewed by an observer. The color depends only on the pitch of the chiral nematic liquid crystal, thus making it quite easy to determine the pitch of a sample placed between crossed polarizers. This effect can produce bright colors, as shown in plate 2.

CONSTRUCTIVE INTERFERENCE IN GENERAL

Constructive interference is also possible even if the light is propagating at an angle to the z-axis. This can be seen from figure 5.7, where the waves emitted from two points in the liquid crystal along the path of the incident wave have been drawn emerging as though they were reflected by the repeating planes in the chiral nematic. To see under what conditions constructive interference occurs, we must calculate how much farther the light reflecting off the lower plane travels relative to light reflecting off the upper plane. If this distance equals one wavelength, then the emerging reflections will be in phase and combine to give constructive interference.

Notice in figure 5.7 that the two paths are equal after point e on the upper path and point c on the lower path. The extra distance light traveling along the lower path encounters is thus equal to distance $(ab + bc)$ minus distance ae. To determine this extra distance, we extend the lower exit path backward and drop a perpendicular to this extension from point a. The two distances ab and bd are equal. This means that the extra path distance $(ab + bc - ae)$ equals $(cd - ae)$, which is just the distance df. But df is a side of the right triangle adf, which has the hypotenuse ad (equal to P) and an angle θ. Therefore, df is just equal to $P \cos \theta$. If the wavelength of the light in the liquid crystal is equal to $P \cos \theta$, then constructive interference occurs

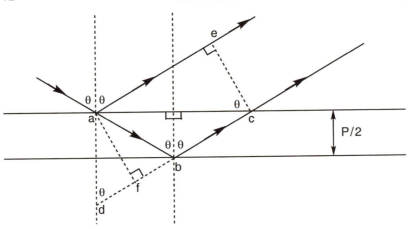

Fig. 5.7. Constructive interference in a chiral nematic liquid crystal when the light is incident at an angle θ with the perpendicular to the identical planes. The lower path is longer than the upper path by $P \cos \theta$ (see text), so constructive interference occurs when the wavelength of the light in the liquid crystal equals $P \cos \theta$.

for light striking the perpendicular to the surface at an angle θ. Notice that this result gives the same answer as before if the light is propagating along the z-axis, since in this case $\theta = 0$ and the condition for reflection simply reduces to the wavelength equaling the pitch.

When chiral nematic liquid crystals are illuminated with white light, it is possible to observe many colors coming from the sample depending on the angle at which the observer views the sample. The longest wavelength reflected comes from the condition that the wavelength in the liquid crystal is equal to the pitch and the reflected light emerges parallel to the z-axis. The wavelength reflected at all other angles is less than this wavelength, with the wavelength decreasing as the angle with the z-axis increases. It is therefore possible for chiral nematic liquid crystals to appear red from one direction but then pass through yellow, green, and so forth as the sample is viewed at larger and larger angles.

THE CANO WEDGE

From our discussion of selective reflection and anomalous optical activity, it is clear that the pitch of a chiral nematic liquid crystal can be measured by observing the wavelength at which selective reflection or anomalous optical activity occurs. One problem with both of these techniques is that the wavelength of light in a liquid crystal is different from the wavelength of the same light in air. The reason for this change is the decrease in the velocity of light when it enters the liquid crystal, a subject covered in chapter 4. Selective reflection or anomalous optical activity occurs when the wavelength of light in the liquid crystal equals the pitch of the chiral nematic. This same light has a longer wavelength in air (by a factor equal to the index of refraction), and this longer wavelength is what one measures in an experiment. Therefore, in order to measure the pitch, the index of refraction of the liquid crystal must be known. This requires an additional experiment.

Fortunately, another method for measuring the pitch exists, which does not require knowledge of the index of refraction. This method involves using two glass surfaces that continuously get farther and farther apart, with a chiral nematic liquid crystal between the two surfaces. One can use two flat glass surfaces to form a wedge, or a lens can simply be placed on top of a flat piece of glass. Either of these arrangements is called a *Cano wedge*, after the French scientist R. Cano, who first developed this technique in the 1960s. If the chiral nematic is constrained to orient with the pitch axis more or less perpendicular to the two glass surfaces and with parallel directors at the two glass surfaces, then there will be positions in the wedge where an integral number of half turns of the helix fit exactly between the two surfaces. This is shown schematically in figure 5.8. In these regions the pitch of the chiral nematic is equal to its normal value, so the liquid crystal is undistorted. On either side of these regions, the liquid crystal is distorted since the pitch must be more or less in order to fit an integral number of half turns between the two glass surfaces.

Even more important is the fact that roughly midway between these regions the number of half helical turns must change by one. This causes a defect there, which is visible as a sharp line under a polarizing microscope. One therefore sees equally spaced lines in the case of a wedge or unequally spaced circles in the case of a lens on a flat surface. If the distance between these lines is measured, then knowledge of the wedge angle or the lens radius

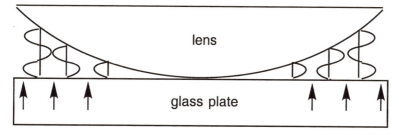

Fig. 5.8. A Cano wedge formed by a lens and flat piece of glass. A chiral nematic liquid crystal between the two glass surfaces is oriented with its helical axis perpendicular to the two surfaces (the curvature of the lens is exaggerated). The helix is shown for positions where integral numbers of half helical turns fit between the surfaces. The vertical arrows show the locations of the defects, which occur at points where the number of helical half turns changes by one.

allows the difference in the distance between the two surfaces at each of the defects to be calculated. This must be equal to one-half pitch.

The color due to selective reflection also changes between the lines, since the pitch slowly changes as the distance between the two surfaces increases or decreases. A beautiful photograph of a Cano wedge formed by using a lens is included as plate 8.

X-RAYS AND SMECTIC LIQUID CRYSTALS

X-rays are electromagnetic waves with a wavelength much shorter than visible light. Typically, X-rays have a wavelength equal to about 0.1 nanometer. Although chiral nematic liquid crystals do not reflect X-rays, because the pitch is so much larger than the wavelengths of X-rays, constructive interference of X-rays does occur in smectic liquid crystals.

Smectic liquid crystals possess a layer structure, so here again there exist regularly spaced regions where the electric and magnetic properties of the liquid crystal are the same. In smectics, the distance between these regions is equal to the separation between layers. This distance is much smaller than the pitch of a chiral nematic, because the layer spacing in smectics is roughly equal to the length of the liquid crystal molecule. Molecular lengths are typically around 3 nanometers.

Figure 5.9 illustrates constructive interference from the layer structure of smectic liquid crystals. Notice how similar this diagram is to figure 5.7. We need perform no additional analysis, since all results remain the same. The

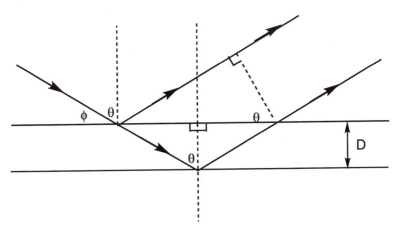

Fig. 5.9. Constructive interference in smectic liquid crystals when the X-rays are incident at an angle ϕ with the layers. The lower path is longer than the upper path by $2D \sin \phi$ (see text), so constructive interference occurs when the wavelength of the X-rays equals $2D \sin \phi$.

condition for reflected X-rays is therefore $2D \cos \theta = \lambda$, where D is the distance between layers. The factor of two results from the fact that the distance between identical regions is D in smectics, whereas it was $P/2$ in chiral nematics. For X-rays, the angle is usually measured from the X-ray beam to the planes in the smectic liquid crystal. This angle has been labeled ϕ in figure 5.9. Since θ and ϕ are complementary, $\cos \theta = \sin \phi$, so the condition for constructive interference for X-rays is usually written $2D \sin \phi = \lambda$. This is a famous relationship, called *Bragg's law*, which was first formulated to explain the reflection of X-rays from solid crystals.

X-ray experiments are an excellent way to measure the layer spacings in smectic liquid crystals. Let us calculate what the angle typically is for such an experiment. If D equals 3 nanometers and λ is equal to 0.1 nanometer, then the reflection has an angle ϕ equal to 1°. These experiments must be designed to detect X-rays reflected at very small angles.

X-ray experiments on smectic liquid crystals have revealed that many different types exist. For example, some smectics have a layer spacing equal to roughly one and a half molecular lengths. These smectic phases occur for molecules that have a tendency to form molecular pairs in which the molecules align themselves parallel to each other by overlapping to some degree. This pairing is usually due to the existence of a permanent electric dipole parallel to the long axis of the molecule near one end. Figure 5.10(a) shows such pairs of molecules and the *smectic* A_d phase that can result. In some cases, such molecules form another type of smectic phase in which the molecules stack up in layers without overlapping, with the electric dipoles concentrated in the boundary between every other layer. A diagram of this phase, called the *smectic* A_2 phase, is shown in figure 5.10(b). Notice that the structure repeats itself over a distance equal to two molecular lengths. This is quite different from the smectic A phase (called *smectic* A_1 with this nomenclature), in which the molecules within each layer point up or down randomly.

As mentioned in chapter 1, some smectic liquid crystals also possess an arrangement for the molecules within each layer. This produces a repeating structure of much smaller length (roughly a molecular width rather than a molecular length) along a direction parallel to the layers. It is possible to observe X-ray reflections due to these repeating regions and thereby gain knowledge about how the molecules are arranged. These experiments show that various arrangements exist. In one common structure, the molecules are located at the center and vertices of connected hexagons.

CHIRAL SMECTIC C LIQUID CRYSTALS

Usually when a liquid crystal possesses a chiral nematic phase, its smectic phases, if present, are either smectic A or smectic C. In some cases, however,

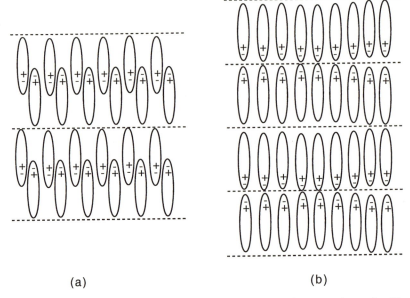

(a)   (b)

Fig. 5.10. The smectic phases formed by molecules that tend to group into pairs. The spacing between the smectic layers is about one and a half molecular lengths in (a), while the distance between identical regions is two molecular lengths in (b). The + and − signs represent permanent electric dipoles at one end of each molecule.

the tendency for the molecule to form a twisted phase is so great that a smectic phase with twist exists, either by itself or along with other smectic phases. One common twisted smectic phase is called the *chiral smectic C* phase or simply the *smectic C** phase. As can be seen from figure 5.11, each layer appears to be typical of a smectic C liquid crystal in that the molecules tend to orient themselves at an angle to the layers, but this direction rotates from one layer to the next. The angle between the director and the smectic layers remains constant, but the director outlines a cone as it rotates around a perpendicular line going from layer to layer. This liquid crystal has a layer spacing about equal to the length of the molecules, but it also possesses a pitch (the distance over which the director makes a complete revolution about the cone). Chiral smectic C liquid crystals therefore reflect X-rays just as other smectic liquid crystals, but they also can reflect light as is typical for chiral nematic liquid crystals. The pitch of these smectic C* liquid crystals tends to be longer than the pitch of chiral nematics, but some still reflect visible light. Smectic C* liquid crystals can also appear colored when viewed under a microscope for the same reason chiral nematics show colors.

Because the molecules in the smectic C* phase are chiral, there is a tendency for one of the two short axes perpendicular to the long axis of the molecules to orient parallel to the layers. If the molecule possesses an electric dipole in this direction, each layer of molecules possesses an electric polarization that rotates as the director rotates around the cone. This electric polarization is denoted by a P in figure 5.11. Because of the electric polarization of each layer, this phase is called a *ferroelectric liquid crystal*. It should be kept in mind, however, that because the electric polarization rotates from one layer to the next, this phase has zero macroscopic electric polarization.

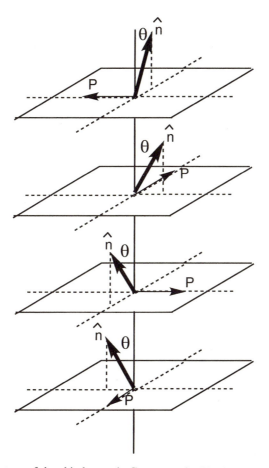

Fig. 5.11. Structure of the chiral smectic C, or smectic C*, phase. The planes represent the smectic layers. The director, $\hat{n}$, always makes the same angle θ with the line perpendicular to the smectic planes, but the orientation of the director rotates around this line in going from one layer to the next. The electric polarization of each layer is given by the arrow labeled P.

There are other compounds in which the director alternates from one side of the cone to the other side in going from one layer to the next. Just as in the smectic C* phase, in moving perpendicular to the layers the director in every other layer rotates around the cone formed by the tilt angle of the director. These phases are called *chiral smectic C$_A$ phases* or *anti-ferroelectric liquid crystals* due to the alternating structure that causes the electric polarization to reverse itself from layer to layer. A slight variation sometimes observed is a phase in which the alternation of the director from one side of the cone to the other is not perfect. This produces a phase with more layers tilted to one side of the cone than the other side and is called a *ferrielectric liquid crystal*.

OTHER TYPES OF SMECTIC LIQUID CRYSTALS

As previously mentioned, unlike smectic A and smectic C liquid crystals, where no positional order exists within the layers, there are smectic phases that display positional order within the layers. The molecules in a *smectic B* liquid crystal tend to be oriented perpendicular to the planes and located at the corners and centers of a network of hexagons. The *smectic F* and *smectic I* phases are tilted analogues of the smectic B phase, in which the tilt of the molecules is perpendicular (smectic F) or parallel (smectic I) to one side of the hexagons. Although there is little correlation between the hexagon networks from one smectic plane to another, the orientation of the hexagons in different planes is consistent. This type of order is called *bond orientational order*.

There are also smectic phases, both tilted and not tilted, that display strong correlations between the preferred positions of the molecules in the different smectic planes. With this much positional order, these phases are more properly called crystals rather than liquid crystals. These include the *crystal B* phase and its tilted analogues the *crystal G* and *crystal J* phases (hexagonal packing), and the *crystal E* phase and its tilted analogues the *crystal H* and *crystal K* phases (rectangular packing).

There is also the *smectic D* phase, which seems to be some type of cubic arrangement of the molecules. There is little evidence for a layered structure, and the phase seems to be isotropic.

Just as the smectic C phase can exist in both an untwisted and twisted form, some of the smectic phases with positional order within the layers also have versions in which the direction of orientational order rotates in going from layer to layer. The *smectic H* and *smectic I* phases are the best examples of this.

Recently, a new type of chiral smectic A phase was reported. In this phase, normal regions with untwisted smectic A layers are separated by defect structures in which the direction of the layers changes abruptly. In going from region to region in a direction perpendicular to the director, the director rotates in a helix (although not continuously). Understanding defects in liquid crystals such as these is important, as we shall see in chapter 10.

Even more recently, new smectic liquid crystal phases formed by *bent-core* or *"banana-shaped" liquid crystal molecules* have been discovered. Figure 5.12 shows a typical bent-core liquid crystal molecule and two of the phases it can form, depicting the molecules as "bows." The direction of the "bows" alternates in going from smectic layer to smectic layer, but the molecules are tilted relative to the perpendicular of the smectic layers. The molecules can all tilt in the same direction or the tilt direction can alternate from layer to layer. The shape of the molecules makes an electric dipole possible only along the direction an "arrow to be shot from the bow" would take. The phase structure therefore lines up the electric dipoles in each layer, producing a liquid crystal phase with *polar order*. Even more interesting, the director (which is perpendicular to the smectic layers), the tilt direction, and the electric dipole are all perpendicular to each other and therefore can form either a right-handed or left-handed arrangement. In the structures shown in

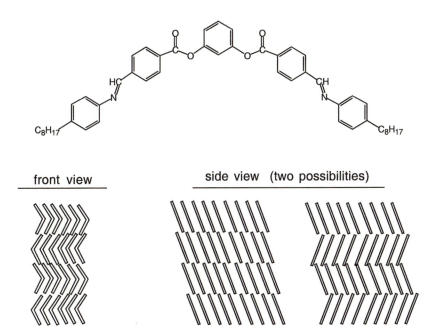

Fig. 5.12. A typical bent-core or "banana-shaped" liquid crystal molecule and some of the phases it forms. The molecules in the front view all tilt in the same direction with respect to the plane of the page (*left side view*) or alternate layer by layer tilting in opposite directions with respect to the plane of the page (*right side view*).

figure 5.12, the arrangement of these three directions alternates handedness in going from one layer to the next. Thus these liquid crystal phases with polar order possess chirality that reverses itself from layer to layer, even though the molecules themselves are nonchiral. As you might expect, application of a strong enough electric field parallel to the smectic layers causes the electric dipoles of all the layers to point in the same direction.

In short, there are many types of smectic phases, which differ in the amount and type of both positional order within the layers and correlation of the order between the layers. The existence of all these different phases offers scientists the chance to examine how cooperative systems behave under varying conditions. For this reason, theoretical and experimental work on these smectic phases is an active area of research.

FREESTANDING LIQUID CRYSTAL FILMS

It is possible to make films using smectic liquid crystals in much the same way soap films are made. A small amount of liquid crystal is placed around a hole in a glass slide, and the side of another glass slide is then drawn across the hole. Some of the liquid crystal is carried across the hole as the second glass slide is drawn, creating a thin film across the hole. The amount of liquid crystal material placed around the hole and the rate at which the second glass slide is drawn determine the thickness of the liquid crystal film. Films consisting of only a few smectic layers can be made in this way. These films are too thin to create interference between the light reflected from the two surfaces, so they appear gray. Thicker films, however, do cause interference at certain visible wavelengths, so these can be highly colored. In general, uniform freestanding films with areas of 1 square centimeter and that are stable for long periods of time can routinely be fabricated.

Liquid crystal films are mostly used for scientific investigations. When only a few layers thick, they represent a system that is virtually two-dimensional in character. If our general understanding of phase transitions is valid, then phase transitions in two-dimensional systems should be quite different from phase transitions in three-dimensional systems. Results from experiments on liquid crystal films support this idea. In addition, close examination of liquid crystal films near phase transitions reveals that in some cases individual layers undergo a transition (e.g., from one smectic phase to another smectic phase) at slightly different temperatures. As a thin liquid crystal film is slowly cooled, for instance, the two outermost layers undergo a phase transition first, and at a slightly lower temperature the two layers just inside the outermost layers change phase. This process continues until at some point the remaining interior layers undergo a phase transition together.

IMAGING LIQUID CRYSTAL MOLECULES

Although the amount of evidence for the fact that liquid crystals involve the orientational order of linear molecules is overwhelming, there is nothing quite like seeing a picture of real liquid crystal molecules showing clearly how they tend to point in the same direction. Likewise, a picture showing the layered arrangement of molecules in a smectic liquid crystal would be extremely convincing. Until recently, such a thought was only a dream, but the development of the scanning tunneling microscope (STM) has changed that forever.

The STM works by measuring the electrical current that passes from a substrate to a tiny metal tip held about 1 nanometer above the substrate. The amount of current (or the height of the tip above the substrate required to maintain a certain current) depends on the atom sitting right below the tip on the surface of the substrate. As the tip is moved across the substrate, the current (or height) is measured at very closely spaced points. By combining many scans, a two-dimensional image can be formed, where the brightness at any point corresponds to the amount of current (or height of the tip) at that point. Such a technique is capable of creating images of individual atoms on the surface of the substrate.

Researchers have performed an experiment in which they placed a small amount of liquid crystal on an extremely flat crystal of graphite. The metal tip of the STM was placed in the liquid crystal and lowered closer and closer to the graphite surface. If the tip was placed close enough to the graphite surface to create an image, but not so close that an image of the graphite surface resulted, then an image of the atoms of the liquid crystal molecules was obtained. These images clearly show that the molecules are oriented parallel to each other, and that the molecules in smectic liquid crystals form layers. As amazing as these images are, there is one drawback to the technique. The molecules being imaged are very close to the graphite substrate, and as previously mentioned, nearby solid surfaces tend to exert strong forces on liquid crystal molecules. As a result, the molecules in the STM images show much more order than occurs in a bulk sample, and also orient along directions defined by the atoms in the nearby graphite surface. In any case, these images are a fitting tribute to the numerous scientists of earlier years who utilized only skilled judgment and intellectual creativity to form our present picture of the various liquid crystal phases.

Chapter 6

LIQUID CRYSTAL DISPLAYS

WE LIVE IN an age of information. This statement usually refers to the vast amount of data available due to the recent explosion in computer technology. There is another aspect to this development, however, which concerns the transfer of this information from machine to human being. For example, most computers must be equipped with an alphanumeric and/or graphics display in order for the user to benefit from the capabilities of the computer. The dedicated microprocessors one finds in automobiles, audio equipment, and aircraft also manage data that must be displayed for the user. Small alphanumeric displays are usually used to convey this information. It is therefore not surprising that rapid development in display technology has accompanied the advances made in computer technology. Liquid crystal displays (LCDs) have played an important role in this development, and appear to be destined to play even a wider role in the future. But before I can discuss how liquid crystal displays function, a brief introduction to the general field of information displays must first be presented.

INFORMATION DISPLAYS

All information displays utilize the ability to control light in order to function. By controlling what parts of a display are bright and what parts are dark, information is passed to the user. The most simple example of how this can be done is the seven-segment numeric display. Light from the area of each of the seven segments is controlled independently, thereby creating the possibility of producing each of the ten digits. Such a display is shown in figure 6.1(a). If both letters and digits must be displayed, a five-by-seven dot matrix can be used. Here light from 35 different areas is controlled independently, producing all the letters and digits with more detail, as shown in figure 6.1(b). Notice that this last display requires that considerably more areas be controlled. The epitome of flexibility is a large-scale display where the entire field of view is composed of small areas (called *pixels*), each of which can be independently controlled. Such a display might have 600 rows and 800 columns (in other words 480,000 pixels), with the capability of displaying both alphanumeric and graphical information.

Regardless of the complexity of the display, the basic working principle remains the control of light from small areas of the display. This can be done in two ways. In the first way, each area can be equipped with the ability to

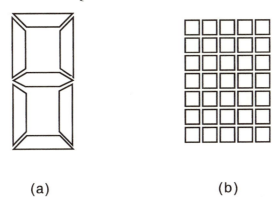

(a)   (b)

Fig. 6.1. Two different display patterns: (a) seven-segment display for numbers and (b) five-by-seven dot matrix display for numbers, letters, and other symbols.

generate light. This is called an *active display*, and two good examples are the cathode ray tube (CRT) and light-emitting diode (LED). The CRT screen contains phosphors that emit light when struck from behind by an electron beam. Different areas are made to produce light by hitting only these areas with the electron beam. In an LED display, each area contains a small light-emitting diode, which gives off light when an electrical voltage is applied to it. By controlling which diodes receive the applied voltage, various letters and numbers can be created. The second way to control light in a display is not to generate light there, but to control the amount of light that passes through or is reflected by the display. This is called a *passive display*. In some passive displays the light is generated by a device behind or beside the actual display; in other cases, ambient light is used. The number of types of passive displays is extremely limited, with LCDs being by far the most important.

Both active displays and passive displays that are illuminated from behind or beside must use electrical power to generate light. Passive displays that utilize ambient light do not consume electrical power in order to generate light and therefore require much less power to operate. This is the largest advantage of liquid crystal displays that rely on ambient light for their operation. The amount of power needed to operate these displays is much less than for other displays, making LCDs ideal for battery-operated equipment, such as wristwatches, portable radios and tape recorders, and pocket calculators.

With both active and passive displays, the control of light from each area is usually achieved by applying a voltage to a device in that area (the CRT is an obvious exception). How well the device functions as a display depends on how it responds to this applied voltage. For example, figure 6.2 shows how the light from a device might depend on the value of the applied volt-

Fig. 6.2. Characteristics of a typical display. The threshold voltage V_{th}, 10% brightness voltage V_{10}, and 90% brightness voltage V_{90} are indicated.

age. There are two characteristics of the response that prove to be important. First of all, how much voltage is necessary to get a change in the brightness? This is called the *threshold voltage* and is indicated in the figure. The second characteristic is how much of a voltage increase is necessary to make the response of the device go from 10% brightness to 90% brightness. This is usually denoted by $V_{90}-V_{10}$; the smaller this quantity, the "sharper" the response. Although a "sharp" response (small $V_{90}-V_{10}$) is very advantageous in displays that are designed to have each area either bright or dark, there is another class of displays that utilize a *gray scale*. In this type of display, each area has the capability of producing various degrees of brightness. This can be accomplished by applying slightly different voltage values between V_{10} and V_{90}, and is obviously easier if the response of the device is not too "sharp."

Another characteristic of any display device is how quickly the device responds as the voltage is either applied or removed. An example of such a response is shown in figure 6.3. Notice that the response to applying a voltage versus removing the voltage may be different. The *switching time*, T_{ON}, is equal to the time between application of the voltage and a 90% response. Likewise the switching time T_{OFF} is equal to the time between removal of the voltage and a 90% change. In most liquid crystal displays, T_{ON} and T_{OFF} differ because the device turns on by responding to the application of a voltage, but simply relaxes back to its off condition when the voltage is removed. The characteristic of the liquid crystal that is most important in determining how quickly it responds is its *orientational viscosity*, that is,

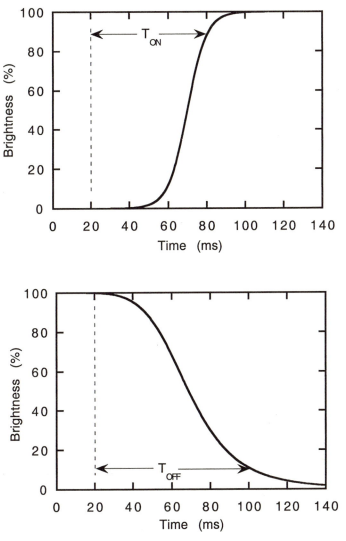

Fig. 6.3. Turn-on and turn-off characteristics for a typical liquid crystal display. T_{ON} and T_{OFF} are the times between application of the voltage change and a 90% change in brightness.

the amount of resistance in the liquid crystal when it is forced to change its director configuration. The time it takes the liquid crystal to relax after the applied voltage is removed strongly depends on its viscosity. Likewise, the time to switch on when the voltage is applied is also affected by the viscosity, but to a lesser degree.

Perhaps one of the most important characteristics of a display is the difference in brightness between an area turned on and an area turned off. In figures 6.2 and 6.3, the brightness in each state is indicated by the height of the curve above the voltage or time axis. This difference in brightness is loosely called the *contrast* of the device; specifically, it is defined as the difference in brightness divided by the larger of the two brightness values. Obviously, the closer the contrast is to 1, the better. Another term often used in describing liquid crystal displays is *contrast ratio*. This is defined as the larger of the two brightness values divided by the smaller of the two values. Devices with good contrast have contrast ratios much greater than 1; contrast ratios are usually in the range from 10 to 100.

An interesting feature of some displays is that the threshold voltage and contrast ratio depend on the angle at which the display is viewed. This is especially true of passive displays, and we shall see that this is a disadvantage in many LCDs.

ADDRESSING DISPLAYS

We must discuss how the voltages to the various devices that make up a display can be applied. If each area of a display is directly connected to circuitry that applies a voltage, we call this scheme *direct addressing*. The advantage is that each area is under constant control, but the disadvantage is that individual connections must be made to each area. Another method that can be used is called *multiplexing*. One example of this kind of addressing is a display where all the areas of a row are connected together and all areas of a column are connected together. This configuration is shown in figure 6.4 for both a seven-segment display and a large-scale graphics display. An individual pixel is turned on by making sure the proper row and proper column receive an appropriate voltage at the same time. Obviously, other pixels in the same row or same column receive some sort of applied voltage, but it must be low enough so as not to turn these pixels on. The obvious advantage of this arrangement is that far fewer connections to the display are required.

In addition, multiplexing does not allow you to apply the proper voltages to all areas at the same time. You might apply a voltage to row 1 while simultaneously applying the proper voltages to each of the columns. This would make all areas of the first row light up properly. Then you would apply a voltage to row 2 and simultaneously apply the proper voltages to each of the columns. This would make all areas of the second row light up correctly. In other words, you would have to cycle through all N rows one at a time, changing the column voltages each time. This means that any area that is on is given the proper voltage only once in each cycle of N rows, and receives smaller changing voltages during the rest of the time. Any area that

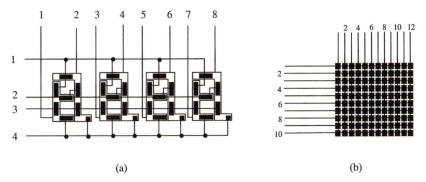

Fig. 6.4. Two types of multiplexed displays. In (a), each row (1–4) is connected to two segments in each digit, and each column (1–8) is connected to four segments of a single digit. Application of a voltage to a row together with the desired voltages on each column turns on two segments in each digit as desired. The same procedure turns on the desired pixels in (b), the matrix display. In a liquid crystal display, the rows are connected to an electrode on one surface while the columns are connected to an electrode on the other surface, with the liquid crystal between the two electrodes.

is off receives a smaller changing voltage all the time. The fraction of time an individual area receives the "on" voltage is called the *duty cycle* and is equal to $1/N$. Liquid crystal displays respond to the average voltage applied over a cycle. Making the applied voltage large when an area is to be turned on is one way to make the average voltage over a cycle for on areas larger than for off areas. Unfortunately, the larger N is, the smaller the fraction of time this larger voltage is applied, and the smaller the difference in average voltage for on and off areas. There is therefore a limit on the number of rows, since the difference in average voltage (and thus the contrast of the display) decreases as the number of rows is increased. A "sharp" voltage response (small $V_{90}-V_{10}$) is the only remedy for this situation, since it might be possible to have the average voltage applied to off areas fall just below the threshold voltage while the average voltage to the on areas (which is only slightly higher if N is large) falls somewhere around V_{90}. In this way good contrast can be achieved with a high *degree of multiplexing, N*.

One last comment should be made about multiplexing. It takes a small amount of time to apply the voltages to a single row and all the columns, so the time required for a full cycle through all the rows increases as N increases. If the duration becomes long enough, areas that are on will appear to flicker. It is therefore advantageous for multiplexed displays to have a somewhat long switching time when an area goes from on to off. This will keep the area on during part of the cycle and lessen the amount of flicker. Even better would be a display that remained on after the voltage was removed until being turned off by another value of the applied voltage.

Such a display is called a *storage display* and can be constructed using liquid crystals.

DYNAMIC SCATTERING MODE

The first liquid crystal displays utilized the light-scattering properties of liquid crystal films when subjected to a large electric field. As discussed in chapter 3, this situation occurs when a liquid crystal that prefers to align with the director perpendicular to the electric field adopts a homeotropic texture between two pieces of glass. Application of a strong electric field causes charged impurities in the liquid crystal to flow, causing turbulence in the film. Although the unperturbed liquid crystal film is clear, the turbulent film intensely scatters light and therefore appears white. A reflector behind the display allows it to function using room light. Such a display is shown in figure 6.5. Evident from the figure is the fact that voltages in LCDs must often be applied to the cell by a *transparent electrode*. This is simply a layer of indium–tin oxide coated on the glass. The layer is thick enough to act as an electrical conductor but thin enough to allow most of the light to pass through. These transparent electrodes can be made in any size or shape, thereby creating the different areas for forming numbers, letters, or graphical information.

The voltage necessary to run a dynamic scattering mode LCD is about 10–20 V, which is slightly higher than for other LCDs. The liquid crystal must possess some charged impurities; this can be achieved by adding impurities to the liquid crystal, or in some cases the liquid crystal ordinarily contains sufficient charged impurities. In a dynamic scattering mode LCD, as in almost all LCDs, the voltages applied to the device are constantly

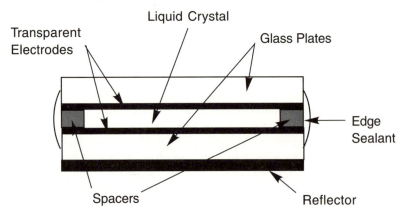

Fig. 6.5. Typical liquid crystal display cell. The reflector is present only if the cell is to be used in the reflection mode.

reversed. This prevents both the buildup of charge on the electrodes and certain electrochemical reactions involving the liquid crystal. Since the dynamic scattering mode depends on the movement of charged carriers, the voltage cannot be reversed too many times each second, or else the charges will not have enough time between reversals to flow very far. Reversing rates up to about 1,000 times per second work well.

The figures in a dynamic scattering mode LCD appear cloudy white on a clear background. Although colored filters can be added to the display, which improves the readability slightly, the contrast is still not very high. For this reason, dynamic scattering mode LCDs are seldom used anymore.

CHIRAL NEMATIC MODE

A storage display can be constructed by utilizing the dynamic scattering mode in a chiral nematic liquid crystal. If the molecules are forced to lie parallel to the glass surfaces, the liquid crystal adopts a texture with the helical axis perpendicular to the glass surfaces. In this state the cell appears clear. If the liquid crystal prefers to align the helical axis perpendicular to the electric field, the application of a fairly large voltage will cause the clear texture to appear cloudy, due to the fact that the texture breaks up into many small unaligned domains. This brightly scattering state remains even after the voltage is removed. An even higher and quickly reversing voltage restores the cell to its clear, aligned state. This type of LCD requires even higher voltages (40–100 V to turn it on and nearly 300 V to turn it off). The switching times are also fairly long (for display purposes), with $T_{ON} = 0.08$ second and $T_{OFF} = 1$ second. Like the dynamic scattering mode LCD, its contrast is not very good.

A different type of LCD can be constructed by using a chiral nematic liquid crystal composed of molecules that prefer to align parallel to an electric field. If the cell is correctly prepared, the liquid crystal adopts an unaligned texture (consisting of many small domains) that appears bright white. Application of a voltage forces the molecules to point perpendicular to the glass surfaces, producing a uniform texture that appears clear. There is no storage capability, however, as the liquid crystal relaxes back to the unaligned, bright texture as soon as the voltage is removed. Although the switching times are much shorter, the contrast is still quite poor.

TWISTED NEMATIC MODE

The twisted nematic mode LCD was developed in the early 1970s and quickly became the workhorse of LCD technology. It remains the predominant type of LCD in use today, mainly because the last twenty years have brought continued improvement in its characteristics.

The *twisted nematic* (TN) display differs from the ones discussed so far by the addition of crossed polarizers to the outside of each piece of glass. As shown in figure 6.6, light incident on the cell from above is polarized along the right/left direction. In addition, the glass surfaces have been treated so the liquid crystal molecules prefer to lie parallel to the surface, but in the right or left direction for the top piece of glass and in or out of the page for the bottom piece of glass. The director of the nematic liquid crystal is therefore forced to twist through an angle of 90° within the cell. This twist is just like the twist of a chiral nematic liquid crystal, so the optical activity in the liquid crystal produces rotation of the polarization direction as light propagates through the cell. With the correct choice of cell thickness and liquid crystal material, the rotation of the polarization direction can be made to follow the twist of the director. The polarization direction of the light is therefore rotated 90° when it strikes the second polarizer. This polarizer allows light with an electric field directed into or out of the page to pass through, which is exactly the state of the light when it reaches the second polarizer. Behind the second polarizer is a reflector that causes the light to traverse back through the cell. Its polarization direction is still into and out of the page, so it passes through the bottom polarizer; is rotated 90° by the liquid crystal again, so it passes through the top polarizer; and therefore emerges from the cell. In this state the cell appears silvery clear.

The situation is quite different when voltage is applied to the cell. The nematic liquid crystal must prefer to align parallel to the electric field for the cell to work. If the applied voltage is above the threshold value for the Freedericksz transition (see chapter 3), the director configuration changes

Fig. 6.6. Twisted nematic liquid crystal display. The polarizers on either side of the cell are crossed. In the absence of an electric field (a), the surfaces induce a 90° twist of the director. This rotates the polarization axis of the light by 90°, thus allowing it to pass through the second polarizer. The presence of an electric field (b) removes this 90° rotation, therefore allowing no light to pass through the second polarizer.

from one of twist to the deformed state shown in figure 6.6(b). Only a small amount of twist remains in the regions next to the glass surfaces. The polarization direction of light traversing the cell is rotated only slightly, meaning that almost all of the light that passes through the top polarizer cannot get through the bottom polarizer. Since no light is reflected back out of the cell, areas with an applied voltage appear dark against those areas without an applied voltage, which appear silvery clear. Removal of the voltage from any area causes immediate relaxation back to the 90° twisted texture. The result is a display with much better contrast and acceptable switching times.

A twisted nematic LCD can be operated in either of two modes. Figure 6.6 depicts a twisted nematic cell that is designed to utilize ambient light. The cell and reflector control the proportion of this ambient light that is redirected back to the viewer's eyes. A twisted nematic cell can also be designed to work in a transmissive mode. All that is necessary is to remove the reflector and place a source of light behind the cell. Light from the source is polarized by the rear polarizer, rotated 90° by the cell if no voltage is applied, and thus passes through the front polarizer. If a voltage is applied, however, the light polarized by the rear polarizer is not rotated by the cell and is therefore blocked by the front polarizer. The viewer sees dark areas against a bright background. Sometimes a transmissive LCD is made with the front and back polarizers parallel to each other. In this case little light gets through the cell if no voltage is applied, while much of the light is transmitted when an applied voltage is present. The viewer therefore sees bright characters on a dark background. Of course, when used in the transmissive mode, additional electrical power must be supplied to generate the light at the source.

One advantage of the twisted nematic LCD is that the threshold voltage is low, on the order of a few volts. However, its greatest advantage is its higher contrast. Unfortunately, this is achieved through the use of polarizers, which introduce problems of their own. For example, polarizers transmit between a third and half of the light incident on them. Using two polarizers therefore reduces the brightness of the display considerably. In addition, the optical activity of the cell and the polarizers themselves work differently depending on the angle at which the light is traversing the cell. Therefore, twisted nematic LCDs generally work best when viewed straight on, with serious degradation of performance as the viewing angle is changed. In addition, the "sharpness" of the brightness versus voltage curve is not great, meaning that a high degree of multiplexing becomes difficult with twisted nematic LCDs. Finally, the switching times for this type of LCD are not particularly fast (0.02 to 0.05 second), so applications requiring a fast response are not feasible. These last two characteristics are the reason that liquid crystal television screens using the twisted nematic mode have not been widely developed.

One problem encountered early on with twisted nematic LCDs was that different areas of the cell would twist in opposite directions when no voltage was applied. This gives the cell a patchy appearance and can significantly degrade its performance. The solution to this problem was to add a small amount of a chiral nematic liquid crystal to the mixture placed in the cell. A chiral nematic liquid crystal twists only in one direction (right- or left-handed), so it forces all parts of the cell to twist in the same direction.

In 1985 a slightly different version of the twisted nematic LCD was introduced. Instead of a 90° twist, the director of the liquid crystal in this new type of cell rotates by 270° in going from one glass surface to the other. The principle of operation remains much the same, because a 270° rotation of the polarization direction of the light produces almost identical effects as a 90° rotation. The advantages of this *super twisted nematic* (STN) display are a slightly "sharper" brightness versus voltage characteristic and less of a problem when the viewing angle is changed. The disadvantage is that the higher degree of twist accentuates the fact that the display affects different wavelengths of light (colors) in slightly different ways, making it more difficult for the off state to appear completely dark. One attempt to rectify this last problem was to prepare *double super twisted* (DST) cells, with one cell right on top of the other. The twist of the two cells is opposite, and only one of the cells is switched on and off by an applied voltage. Since the unswitched cell is always in its 270° twist state, it compensates for any wavelength dependent effects in the switched cell when the switched cell is in the off state. The result is a darker off state and therefore an increase in contrast. Additional modifications of the STN display were the *super birefringence twisted effect* (SBE) cell and the *optical mode interference effect* (OMI) display. Both of these utilize large twist angles.

Twisted nematic LCDs can be made colored by the addition of colored filters or colored polarizers. If a single color is used, the display appears brightly colored on a black background or vice versa. If two differently colored polarizers are placed at right angles in front of the cell with a normal polarizer behind the cell, the characters will appear in one color and the background in another color. The reason for this is that certain colored polarizers effectively polarize light only of one color, while light of other colors is passed without being polarized. This results in light of two colors with polarization directions at right angles entering the cell. Both are rotated by the cell, but only one is passed by the normal polarizer, which one depending on whether the cell is on or off. In either case, only one color of light emerges from the cell. Some of these colored twisted nematic LCDs are quite appealing.

An interesting possibility in designing twisted nematic LCDs concerns the fact that some liquid crystals prefer to align with the director parallel to the electric field if the frequency of the alternating voltage is low, but prefer to align with the director perpendicular to the field if the frequency of the applied voltage is high. The cell can therefore be switched to its twisted state by a high-frequency voltage and switched to its untwisted state by a low-frequency voltage. This method of operation is called *dual frequency addressing*. The advantage of this technique is that the switching times are reduced, but the use of high frequencies results in higher operating voltages and an increase in the power consumed.

LIQUID CRYSTAL DISPLAYS USING DYES

A dye is a substance that absorbs light of a certain wavelength, causing the light reflected from or transmitted through the dye to appear colored. Some dye molecules absorb light of a certain wavelength better if the light is polarized along one axis of the molecule. Such dyes are called *dichroic dyes*, and these are the ones used in liquid crystal displays.

When dye molecules that are elongated in shape are dissolved in a liquid crystal, they tend to orient along the director of the liquid crystal. This effect is called the *guest-host interaction*, in that the dye molecules are the "guest" molecules being oriented by the liquid crystal "host" molecules. The dichroic properties of certain dye molecules can be utilized for display purposes by applying electric fields to the liquid crystal and causing reorientation of both the liquid crystal and the dye molecules.

Consider the display cell shown in figure 6.7. Light passing through a single polarizer on the front of the cell polarizes the light along the director of a liquid crystal that contains some dissolved dye aligned homogeneously. If the dye absorbs better when the light is polarized along the long axis of the dye molecules, the off state shown in figure 6.7(a) appears brightly colored in transmission or reflection. If the liquid crystal prefers to orient along the electric field direction, the on state shown in figure 6.7(b) shows much less color in transmission or reflection, since the light is polarized perpendicular to

(a) (b)

Fig. 6.7. Guest-host liquid crystal display using a dichroic dye. An electric field (b) causes a reorientation of the liquid crystal, which in turn reorients the dye molecules. The absorption characteristics of the cell therefore change from the condition with no applied electric field (a).

the long axis of most of the dye molecules. The result is colorless characters on a colored background. Just the opposite effect occurs if the liquid crystal is aligned homeotropically and the liquid crystal molecules prefer to orient perpendicular to the electric field. This results in colored characters on a colorless background.

The advantage of the guest-host LCD stems from the fact that only one polarizer is present. This increases the brightness of the display and reduces the change in contrast as the viewing angle is varied.

A slightly different version of the guest-host LCD operates with no polarizers. A chiral nematic liquid crystal with a fairly short pitch acts as the host material, and a dichroic dye is dissolved in it. The surfaces are treated so that the sample is composed of many domains, each with a helical axis pointing in some direction, as shown in figure 6.8(a). The shortness of the pitch means that the director in each domain rotates around many times. This state is highly colored, because no matter what the polarization of the incoming light is, some of the dye molecules are oriented with their long axes parallel to the polarization direction. Assuming the liquid crystal molecules prefer to orient themselves parallel to an electric field, application of a voltage causes the dye molecules to lie perpendicular to the polarization direction for the incoming light. As evident from figure 6.8(b), this is actually a voltage-induced chiral nematic to nematic phase transition. The result is a colorless on state against a brightly colored off state. Again, this type of display can be used in transmission or reflection. The lack of any polarizers makes this a very bright display, with superior characteristics over a wide range of viewing angles. In addition, the cell operates with a faster switching time in going from the on to the off state, since the twisting nature of the off state is due to the fact that a chiral nematic material is used and not simply due to surface alignment effects.

Fig. 6.8. Guest-host dye LCD utilizing a chiral nematic phase transition due to an applied voltage. The absorption characteristics of the dichroic dye molecules are different for the two phases.

BIREFRINGENT LIQUID CRYSTAL DISPLAYS

In my discussion of birefringence, I explained how the phase angle between light polarized along one direction and light polarized along a perpendicular direction changes continuously as the light propagates through a liquid crystal. This effect can also be used to make a liquid crystal display.

Consider the situation shown in figure 6.9. The two polarizers are oriented at an angle of 45° to the director of the liquid crystal in the off state. As before, we therefore can think of the light coming out of the first polarizer as having two components: one polarized along the director of the liquid crystal and the other polarized perpendicular to the director. As the light enters the liquid crystal, these two polarizations have a phase angle between them of zero. As they propagate through the cell, the phase angle increases, so at times the light is circularly polarized, linearly polarized, or elliptically polarized. If the thickness of the cell is just right, the light coming out of the cell is linearly polarized along the axis of the second polarizer (which is shown oriented perpendicular to the first polarizer). The light passes through, and the display appears bright. If the liquid crystal material tends to line up with an electric field, the on state (figure 6.9[b]) shows no birefringence, since the light polarized in any direction perpendicular to the director travels at the same velocity. There is no phase change, so no light passes through the second polarizer. The result would be dark characters on a bright background, whether the cell is used in transmission or reflection.

Such cells are called *selective interference liquid crystal displays* because the phase angle of the emerging light strongly depends on wavelength. The

Fig. 6.9. A selective interference liquid crystal display. The polarizers are crossed, and the director in the off state makes an angle of 45° with the polarizers' axes. The phase angle change caused by the liquid crystal in the off state (a) allows light to pass through the second polarizer. The on state (b) produces little phase angle change, so little light passes through the second polarizer.

off state is therefore brightly colored, since the phase angle is correct for only one wavelength. Unfortunately, this color depends on the temperature, viewing angle, and thickness of the cell. Two advantages of this display are that the voltage threshold is "sharper" and the switching times are somewhat faster than in the twisted nematic LCD.

A slight variation of this display utilizes a liquid crystal that tends to orient perpendicular to an electric field. In this case the cell is always birefringent, but an applied voltage changes the phase angle of the emerging light by causing the order parameter of the liquid crystal to vary slightly. This changing phase angle produces variation in the color of light that is allowed to pass through the second polarizer. Large changes in voltage produce two distinct colors (an off and on state), but continuously variable colors are also possible for intermediate values of the applied voltage. This *tunable birefringence liquid crystal display* is therefore capable of producing beautiful and continuously variable colors, which make it perfect for some projection-type displays.

To avoid the problems of color uniformity due to variations in the thickness of the cell, viewing angle changes, and temperature drifts, the birefringent effect can be combined with a twisted nematic LCD. A sheet of solid birefringent material of uniform thickness (calcite, for example) is placed between the liquid crystal and the polarizer of a twisted nematic cell (see figure 6.10). Light from above, after traversing the twisted nematic cell, is polarized either parallel or perpendicular to the initial polarization direction depending on whether the cell is off or on. In either case, this direction is at 45° to the axes of the solid retardation sheet. The retardation plate introduces a phase shift of 180°, which produces light polarized at an angle of 90° relative to the light that entered the retardation plate. This light either

Fig. 6.10. LCD with a solid birefringent plate. The twisted nematic LCD either rotates the plane of polarization by 90° (a) or does not rotate the plane of polarization (b). The solid retardation plate causes a phase angle change, which does or does not allow the light to pass through the second polarizer.

passes through the second polarizer or does not, depending on whether the twisted nematic cell rotated the light from the first polarizer or not. Actually, the retardation plate works this way for only one wavelength (color) of light, so one sees one color in the off state and the complementary color in the on state. This display is therefore a two-color display, with the two colors being complementary. Since the birefringent material is a solid, its thickness is quite uniform, and the phase angle change does not depend strongly on temperature. The viewing angle problem remains, and is in fact slightly worse then in twisted nematic LCDs.

Recently, an entirely new type of birefringent LCD has been commercialized, and this display has the possibility of accelerating significantly the popularity of LCDs. Instead of placing the electrodes on each piece of glass so that the electric field is perpendicular to the glass, in this new display the electrodes are placed on the same piece of glass. For much of the region containing the liquid crystal, therefore, the electric field is parallel to the plane of the glass, opening up an entirely new realm of possibilities.

The basic operation of this new display with *in-plane switching* (IPS) is shown in figure 6.11. In the off state with no voltage applied, surfactants on the glass align the liquid crystal parallel to one of the crossed polarizers. In this state no light is transmitted. When a voltage is applied in the on state, the electric field rotates the director while keeping it parallel to the glass so that it now makes an angle with each of the polarizers. This makes the cell birefringent, changing the linearly polarized light to elliptically polarized light as it passes through the liquid crystal. For maximum transmission, the voltage necessary for an electric field that rotates the director by 45° is used. For less than maximum transmission, and therefore a gray scale, a voltage less than this is used.

Fig. 6.11. LCD employing in-plane switching. The director is always parallel to the glass surfaces, being parallel to one of the polarizers in the off state (a) and making an angle with the polarizers in the on state (b). The curved arrow in (b) represents the electric field direction.

The chief advantage of displays utilizing in-plane switching is that the display works well when viewed from an angle. The reason for this is quite interesting. In a twisted nematic display, the rotation of the linearly polarized light by 90° in the off state, which is necessary for the display to be truly dark, is achieved only for light traveling perpendicular to the glass. When the viewing angle moves away from this direction, the rotation of the light decreases from 90°, and the state becomes less and less dark. This reduces the contrast and causes color changes. When IPS is employed, the off state is truly dark regardless of the angle at which the light strikes the glass. IPS displays still show some variation depending on whether the viewing angle is from the top or bottom as opposed to the side. This problem is solved by fabricating the electrodes for different pixels so that the electric field points in more than one direction while remaining parallel to the glass. This smooths out the variation and produces a display that can be viewed from almost any angle with no degradation in contrast or color. This represents another new idea in liquid crystal display technology. No longer does the liquid crystal in a display have to be a single domain in which the director configuration is the same everywhere. Fabricating the electrodes at different angles, even allowing them to be three-dimensional in some cases, achieves significant improvements in these *multidomain LCDs*. Plate 9 is a picture of a new LCD computer monitor employing these new technologies.

SMECTIC LIQUID CRYSTAL DISPLAYS

Smectic liquid crystals usually do not respond to applied electric and magnetic fields as easily as nematic liquid crystals do. For this reason, simple application of an electric field rarely produces enough change in the optical properties of a smectic liquid crystal to be useful for display purposes. However, the texture of a smectic liquid crystal does depend strongly on whether it cooled to form the smectic phase in the presence or absence of an electric field. For example, a liquid crystal that prefers to align its director along an electric field will cool into a smectic phase with the layers parallel to the glass surfaces if the electric field is present during cooling. This is shown in figure 6.12(a). This texture is very uniform and scatters little light. If no electric field is present during cooling, however, the smectic liquid crystal texture is unaligned, with the smectic planes oriented in different directions in different parts of the sample. This is called the *focal conic texture* and is shown in figure 6.12(b). This texture strongly scatters light, so the cell appears milky white. If dye molecules are present in the smectic liquid crystal, colors other than white can be produced. Both the clear and scattering textures remain after the cooling process, even if the electric field is removed. This means that smectic LCDs do not need to be frequently refreshed, so they are ideal for high degrees of multiplexing. In addition, no polarizers are

OFF   ON

(a)   (b)

Fig. 6.12. Smectic liquid crystal display. If cooled from the nematic phase in the presence of an electric field, the state shown in (a) forms and is transparent. Cooling without an electric field produces the state shown in (b), which is highly scattering.

necessary, so these displays have good contrast and can be viewed at wide angles.

Smectic LCDs have been addressed in two ways. In the first method, a voltage that causes local heating to the higher-temperature liquid crystal phase is applied to a row. Immediately following this heating pulse, voltages are applied to all the columns that should be clear in that row. Pixels with applied voltages cool in the presence of an electric field and appear transparent, while the other pixels of that row cool without an applied electric field and appear colored. The heating pulse is then applied to the next row, followed by the application of voltages to the proper columns. This process continues for all rows of the display and then repeats itself. The second method employs a laser. First, the entire display is quickly heated and then allowed to cool in the presence of an electric field. Second, a laser beam scans across the clear display, heating the liquid crystal wherever it hits. Since the electric field has been removed, the areas heated by the laser beam cool with no electric field applied, and thus appear colored. Although the information content of these laser addressed smectic LCDs can be extremely high, the laser itself and its projection optics are quite expensive. Still, these displays are useful for large-area projection. Whether thermally or laser addressed, the need for electrical power to heat the liquid crystal eliminates one of the chief advantages of most LCDs—low power consumption.

Another type of LCD uses a chiral smectic C liquid crystal instead of a smectic A. As shown in figure 5.11, chiral smectic C liquid crystals possess a director that rotates in a cone in going from one smectic layer to the next. However, if the cell is thin enough, the interaction of the liquid crystal with the surfaces produces a texture in which there is no rotation of the director within the cell. If the director prefers to align parallel to the glass, the two

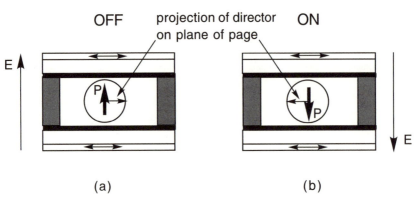

Fig. 6.13. Surface stabilized ferroelectric LCD. The layers of the smectic C* liquid crystal are in the plane of the page. The director is at an angle of 22.5° to the line perpendicular to these planes, and is switched between the two states shown by the action of an electric field on the permanent dipoles (labeled P) across the molecules. The crossed polarizers are parallel and perpendicular to the director in (a), so this state is dark. The director is at an angle of 45° to the polarizer axes in (b), so the birefringence of the liquid crystal causes a 180° phase angle shift. This changes the polarization angle by 90°, making the cell become bright.

states shown in figure 6.13 are possible. In both of the states the smectic layers are in the plane of the page, and the director is parallel to the glass surfaces but pointing at an angle of 22.5° to the line perpendicular to the smectic planes. As mentioned in chapter 5, the fact that this phase prefers to be chiral means that the molecules preferentially orient one of the two short axes perpendicular to the long molecular axis. If the molecules possess a permanent electric dipole moment along this direction, then the liquid crystal itself possesses an electric polarization perpendicular to the director. Since in these cells there is no rotation of the director, the electric polarization of each layer points in the same direction, producing a phase with a macroscopic electric polarization. This cell is called a *surface stabilized ferroelectric liquid crystal* (SSFLC) device.

To understand how this device can switch from one orientation to another, notice that each of the two states shown in figure 6.13 possesses an electric polarization (labeled P) perpendicular to the glass surfaces. If an electric field is applied by transparent electrodes on the glass, the liquid crystal's electric polarization responds by aligning with the field; an electric field in one direction causes the liquid crystal to adopt the state shown in figure 6.13(a), while an electric field in the opposite direction produces the state shown in figure 6.13(b). If crossed polarizers are present and aligned parallel and perpendicular to the director in the state shown in figure 6.13(a), then this state appears dark since the incoming light is polarized along the director and no phase retardation occurs. However, if the tilt angle is 22.5°, the light entering the liquid crystal state shown in figure 6.13(b) is polarized at an angle of 45° to the director. The two components of the light (parallel and perpendicular to the director) therefore suffer a phase shift in passing through the cell. If the thickness of the cell is adjusted so a 180° phase shift results, the emerging light is linearly polarized, but at an angle of 90° to the incoming light. The cell therefore appears bright. The switching times for these displays are faster than for other LCDs by a factor of one hundred to one thousand, because (1) the sample possesses ordered permanent electric dipoles, and (2) an electric field forces both the on and off transitions. Thus ferroelectric LCDs may play a significant role in displays that must have very short response times. In addition, since no heating or laser beam is necessary, these displays use very little power.

As also discussed in chapter 5, in anti-ferroelectric liquid crystals the electric polarization points in opposite directions from one layer to the next. This remains the case when such a liquid crystal is placed in a thin cell and stabilized by surface interactions. With no voltage applied to the cell, there are just as many layers with the electric polarization pointed in one direction as layers with the electric polarization pointed in the other direction. If a voltage is applied with one polarity, the electric polarization of half the layers reverses, so a macroscopic electric polarization appears. The same is true if a voltage with the opposite polarity is applied, but in this case the electric polarization in the other layers reverses to produce a macroscopic electric polarization in the opposite direction. Thus anti-ferroelectric liquid crystals can be switched into one of three states, a property called *tristate switching*. More important, however, is the fact that the switching in anti-ferroelectric liquid crystals appears to occur at a threshold value of the applied voltage, unlike the switching in ferroelectric liquid crystals, which appears to be continuous or thresholdless (occurs gradually as the applied voltage is increased). As noted earlier, a sharp threshold makes multiplexing much easier, so anti-ferroelectric liquid crystals may have significant advantages over ferroelectric liquid crystals for certain display applications.

Before surface stabilized ferroelectric or anti-ferroelectric displays can be exploited commercially, orientation problems of the liquid crystal inside the cell must be solved. The most significant of these is the fact that when such a cell is cooled from the smectic A phase to the chiral smectic C phase, a chevron texture results due to the combined constraints of the surface interactions and the tilt angle. The surface interactions tend to lock the molecules at the surface in place, so a smectic A layer that went perpendicularly from one surface to the other must form a bend at the center once the chiral smectic C phase forms and the layers are tilted with respect to the director. This is shown in figure 6.14, where it is clear that there are

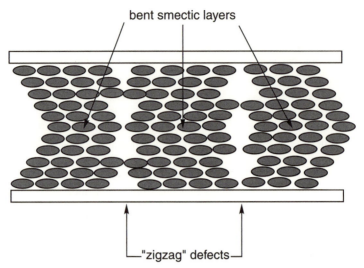

Fig. 6.14. Chevron textures in SSFLCs. Treatment of the surfaces for parallel alignment of the molecules plus the appearance of tilt when the liquid crystal is cooled
from the smectic A to the chiral smectic C phase cause both directions of chevron
bending to occur. "Zigzag" defects separating regions of the two types of textures
degrade the performance of these devices.

two ways for the layers to bend at the center. Since there is nothing to force
one type of bend over the other, a typical display develops regions of both.
Defects that zigzag through the cell are present where these regions meet,
raising the amount of light scattered from the cell and reducing the cell's contrast. Researchers have attempted to overcome this problem in many different
ways, but a successful solution has been achieved only recently. Plate 10
shows an SSFLC display being used in a television. The display is quite
small, and the picture is significantly magnified when it is projected onto a
screen.

LIQUID CRYSTAL DISPLAYS UTILIZING POLYMERS

A recent development in liquid crystal displays utilizes polymers either to
contain the liquid crystal material or to promote a specific texture. These
displays are easily fabricated and offer some significant advantages over other
types of LCDs.

In the first type of LCD utilizing polymers, the liquid crystal material is
contained in tiny droplets embedded in a solid polymer matrix as shown in
figure 6.15(a). One might think this is difficult to achieve, but in fact it is
quite easy. The key is to form the solid polymer by mixing a fluid polymer with a fluid cross-linking agent. The most common example of this is

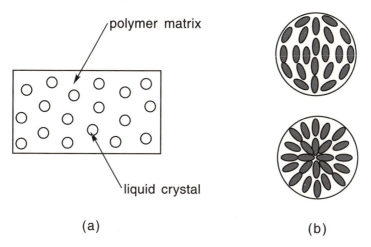

Fig. 6.15. Polymer dispersed liquid crystal display. The droplets of liquid crystal are shown dispersed throughout the polymer in (a). The two possible director configurations inside the droplets are shown in (b).

an epoxy that comes in two parts (resin and hardener). Once mixed, chemical reactions cause cross-links between the normally fluid polymer to form, resulting in a solid material. If a liquid crystal material is dissolved into the two-part mixture while it is still fluid, the cross-linking reactions force the liquid crystal out of the solid polymer matrix and into fairly spherical droplets of almost pure liquid crystals. The size of the droplets is fairly uniform and can be varied from 0.1 to 10 micrometers in diameter by adjusting the rate at which the cross-linking reaction takes place. If the polymer and liquid crystal harden between two parallel surfaces, a *polymer dispersed liquid crystal* (PDLC) film results.

These films normally scatter light extremely efficiently. The reason for this can be seen in figure 6.15(b). The liquid crystal director inside each droplet adopts one of two configurations, depending on whether the liquid crystal prefers to align parallel or perpendicular to the polymer surface. In the case of parallel alignment, two points where the director is undefined (defects) occur, while the case of perpendicular alignment causes only one defect to form in the center of the droplet. Let us concentrate on the case of parallel alignment. If no electric field is applied, each droplet is oriented with the line connecting the two defects oriented completely at random. Remember that light polarized parallel to the director travels at one velocity in the liquid crystal (index of refraction $n_{\parallel}$) while light polarized perpendicular to the director travels with a different velocity (index of refraction $n_{\perp}$).

The solid polymer is isotropic, so it has a single index of refraction (n_S) that in general is different from either of the liquid crystal indexes. Light traveling through the PDLC film has its electric field oriented parallel to the surfaces of the film, so it must interact with droplets oriented with the director parallel to the electric field, perpendicular to the electric field, and every angle in between. It is impossible for the index of the polymer n_S to equal both $n_{\parallel}$ and $n_{\perp}$, so the sudden change of index causes reflections from most of the droplets. As evident from figure 6.16(a), a PDLC film with no electric field applied scatters light intensely.

However, if an electric field is applied across the PDLC film and the liquid crystal molecules prefer to align parallel to the field, the director configuration in each droplet is oriented so the line connecting the two defects lies parallel to the electric field. This is shown in figure 6.16(b). Light passing through the cell now has its electric field nearly perpendicular to the director, so if this index ($n_{\perp}$) is made equal to the polymer index (n_S), little reflection occurs at the boundary of each droplet. The PDLC film appears clear. To optimize the light-scattering properties of these films, researchers carefully select the proper materials, droplet size, and film thickness.

There are two major advantages to these displays. First of all, no polarizers are necessary, so these displays appear quite bright. Second, the thickness of the film is not critical, unlike with most other LCDs. Large-scale displays are easily fabricated and can even be flexible. One chief disadvantage is that the contrast declines for viewing angles away from perpendicular, because light traveling in these directions does not have its electric field perpendicular to the director (for which the index of the polymer is matched). Another disadvantage is that the voltage threshold curve is not sharp, so a high degree

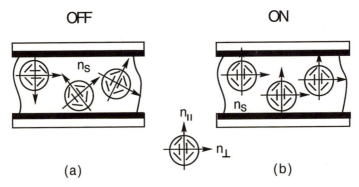

Fig. 6.16. Reorientation of the director within the liquid crystal droplets due to an electric field. The directors inside the droplets are unaligned with no electric field present (a), so mismatch between the droplet index of refraction (namely, $n_{\parallel}$) and the polymer index of refraction n_s scatters light. When the director inside the droplets is aligned by an electric field (b), equal indices of refraction ($n_{\perp}$ and n_s) reduce the scattering of light and produce a clear state.

of multiplexing is difficult with these displays. Perhaps the ease of large-scale fabrication will be the aspect that ensures a market for these PDLC displays. Switchable windows in buildings and automobiles, switchable partitions in office spaces, and large advertising signs (constantly changing!) may become commonplace. Plate 11 is a picture of a switchable window that is currently available. This window is actually manufactured in a two-step process called *microencapsulation*. First, tiny droplets of liquid crystal are surrounded with a thin coating of polymer. Second, these encapsulated droplets are then introduced into a plastic or polymer matrix. Such films have properties very similar to the PDLC films.

In the second type of LCD using polymers, a small amount of polymer is mixed with a chiral nematic liquid crystal. The surfaces of the cell are treated so the liquid crystal molecules orient parallel to the surfaces. If the pitch of the chiral nematic liquid crystal is correct, a planar or Grandjean texture is produced as the cell strongly reflects one color of light. If at this point the cell is exposed to ultraviolet light, the polymer links with itself, forming a network throughout the liquid crystal cell. Because the network is formed while the liquid crystal is in the Grandjean texture, the network serves to stabilize this texture. If the liquid crystal molecules prefer to align along the direction of the electric field, the application of a voltage to the cell serves to break up the Grandjean texture into domains of fingerprint textures. This texture reflects little light, so it appears dark if a light-absorbing material is placed behind the cell. Interestingly enough, if the applied voltage is removed, the fingerprint texture remains. Thus the highly colored state and the dark state are both stable with no applied voltage. If a higher voltage is applied and decreased to zero abruptly, the Grandjean texture results. There are some nice advantages to this *polymer stabilized cholesteric texture* (PSCT) display, including lack of polarizers, on and off states stable in zero electric field, and comparable switching times in both directions.

As evident from the discussion in this chapter, the field of liquid crystal displays is a technology with many possibilities. Over the last twenty years, for example, the market for LCDs has increased by about 25% each year, and there is no reason to think this will slow down significantly in the near future. One interesting aspect of this exponential growth is that early on LCDs did not necessarily replace other displays that worked well in some applications; rather, LCDs were used in new ways and therefore opened new markets. The net result is that at present the market for LCDs is comparable to that of CRTs. Over the past ten years, LCDs became the display of choice for laptop computers. Now LCDs are starting to find their way into the desktop computer monitor market. The monitor shown in plate 9 is an example of a new LCD computer monitor with characteristics that truly rival CRT monitors (it can even be used to view cable television).

The improvement in LCD design over the last ten years has been impressive. A typical thin-film transistor (TFT) liquid crystal display in 1990 consumed 18 watts of power, had a mass of 960 grams, and was 25 millimeters thick. In 2000, a TFT display typically consumed 5 or 6 watts, had a mass of 350–450 grams, and was only 5 millimeters thick. In this same time period, the contrast increased from about 80:1 to around 350:1, the viewing angle increased from roughly 90° to 160°, the response time decreased from approximately 50 milliseconds to close to 20 milliseconds, and the number of pixels increased from a little over 300,000 to nearly 2,000,000. No wonder the global market for LCDs increased from roughly $2 billion to over $20 billion during this decade. In 2000, a little less than half of the LCD market was in laptop displays, a little less than a quarter was in computer monitors, a little over 10% was for cell phone displays, and the remainder was in games, televisions, projectors, and digital instruments.

The next development in the future of LCDs will probably be displays with a significantly faster response and much higher resolution (pixels per inch). Not only might these new displays displace CRTs in televisions, but it is likely that our multi-media displays of the future will utilize LCDs. In addition, there is no reason to think that LCDs will not be the internet terminals and wireless information components we use tomorrow.

Chapter 7

LCD TECHNOLOGY AND OTHER APPLICATIONS

THE POSSIBILITY OF using liquid crystals in various types of displays was successfully demonstrated almost thirty years ago, yet it took ten to twenty years to develop the technology to manufacture LCDs of high quality and reliability. The first part of this chapter discusses the progress that has been made in the field of liquid crystal displays, ending with a description of efforts to utilize semiconductor technology in LCDs. Although displays are the most important application for liquid crystals at this time, other possibilities exist, and some of these have been fully developed. These applications are discussed in the latter part of the chapter, and include descriptions of devices such as light valves and temperature sensors.

MATERIALS

As discussed in chapter 1, molecules that form liquid crystalline phases are elongated and fairly rigid in the central portion of the molecule. This is usually achieved by linking two or more benzene rings together in a linear arrangement. Flexible hydrocarbon chains attached to each end of the central benzene rings are also necessary to form liquid crystalline phases. The linkage group between the benzene rings is probably the most important part of the structure, so scientists usually refer to it by name. Hydrocarbon chains of various lengths can be added to the central portion, resulting in a series of liquid crystalline compounds. These different compounds possess different liquid crystal phases, giving the scientist or engineer many choices of materials for study or use.

One example of a liquid crystal series is based on one of the earliest liquid crystal molecules of significant importance. Figure 7.1 shows the first three members of the *p*-azoxyanisole (PAA) series. The central linkage between the two benzene rings is the azoxy group for all members of the series. Only the numbers of carbon and hydrogen atoms in the hydrocarbon chains on each end of the molecule vary from one member of the series to the next. Both the temperature at which the liquid crystal phases form and the type of liquid crystal phase that forms vary from one member of the series to the next. This is typical of a liquid crystal series.

One way to illustrate this is to make a graph showing the transition temperatures as the number of carbon atoms in the chains is increased by one.

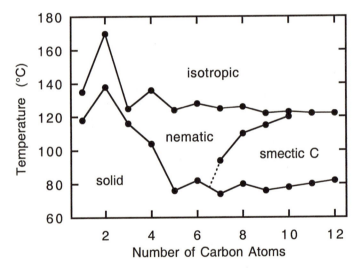

Fig. 7.1. First three members of the *p*-azoxyanisole (PAA) homologous series. The phases and transition temperatures of these compounds are given in the next figure.

Fig. 7.2. The transition temperatures and phases for the first twelve members of the PAA homologous series.

Such a graph for the PAA series is shown in figure 7.2. Two features of this graph are common to many liquid crystal series. First, the temperature at which the liquid crystal phases are stable slowly decreases as the length of the hydrocarbon chains is increased. Second, the smectic phase is more likely to be present if the hydrocarbon chains are longer. Obviously, the presence of long hydrocarbon chains produces interactions between the molecules that promote the formation of layered structures.

In the PAA series, an oxygen atom connects the hydrocarbon chain with the benzene ring. An entire series of liquid-crystal-forming compounds with slightly different properties results when the hydrocarbon chains are connected directly to the benzene rings. This fact is illustrated in figure 7.3, where a graph of the transition temperatures for this series appears. Notice that the transition temperatures are much lower, and that a smectic A phase is present instead of a smectic C phase.

The central portion of the molecule is probably the most important feature in determining what liquid crystalline phases are present. The length of the hydrocarbon chains on either end and how the chains are attached to the central portion of the molecule can then be varied to achieve a wide range of transition temperatures and to some extent change the phases present. Progress in the development of materials suitable for LCDs has therefore come from research into new ways of linking the benzene rings together in the central portion of the molecule.

This point is nicely illustrated by the efforts around 1960 to develop a room temperature liquid crystal. The high transition temperatures of the azoxy and azo series make them unsuitable for display applications, so the search for a series with lower transition temperatures was of utmost importance in the early development of LCDs. The result (MBBA) has already been

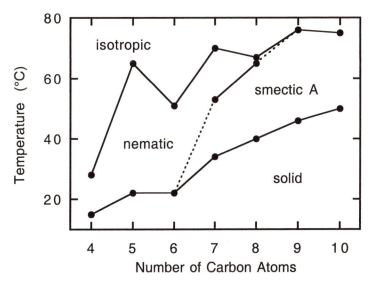

Fig. 7.3. The transition temperatures and phases for the fourth through tenth members of the alkylazoxybenzene homologous series.

mentioned in chapter 2. The members of this series are called Schiff bases because of the central linkage group (see figure 2.2). Notice that MBBA has an oxygen atom linking one hydrocarbon chain to the benzene ring but no oxygen atom linking the other hydrocarbon chain. A series can therefore be formed by adding more carbon and hydrogen atoms to the oxygen-linked chain only. The transition temperatures and phases for this series are shown in figure 7.4. Note how low the transition temperatures are. Another series can be made by adding more carbon and hydrogen atoms only to the chain that is not connected to the benzene ring via an oxygen atom. Figure 7.5 shows the transition temperatures and phases for this series. The liquid crystalline phases occur at low temperatures in this series also.

One other central linkage group that was studied for use during the early development of LCD technology is the ester formed from benzoic acid. Two examples of these compounds are shown in figure 7.6. The transition temperatures of these compounds are fairly low, with the sequence of phases being different among members of the series.

MODERN MATERIALS

One of the most perplexing problems that haunted the development of materials suitable for LCDs was the lack of chemically stable liquid crystals.

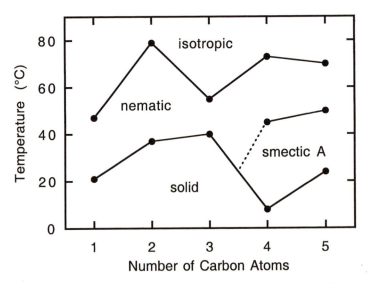

Fig. 7.4. The transition temperatures and phases for the first through fifth members of the Schiff base homologous series created by changing the length of the oxygen-containing chain only. The compound with one carbon is *p*-methoxybenzylidene-*p*-*n*-butylaniline (MBBA).

Plate 1. Nematic liquid crystal. The liquid crystal sample is between crossed polarizers. The liquid crystal appears dark where the director points along one of the polarizer axes. Points where dark areas converge are disclinations. (Courtesy of E. Merck Company)

Plate 2. Chiral nematic liquid crystal (Grandjean texture). The liquid crystal sample is between crossed polarizers, and the helical axis is perpendicular to the picture. The white lines running through the blue background are disclinations. (Courtesy of H. Kitzerow)

Plate 3. Chiral nematic liquid crystal (fingerprint texture). The sample is between crossed polarizers, and the helical axis is in the plane of the picture. The lines are individual turns of the helix. (Courtesy of H. Kitzerow)

Plate 4. Smectic A liquid crystal. The liquid crystal sample is between crossed polarizers. (Courtesy of M. Neubert)

Plate 5. Smectic B liquid crystal. The liquid crystal sample is between crossed polarizers. (Courtesy of J. Goodby)

Plate 6. Discotic liquid crystal. The liquid crystal sample is between crossed polarizers. The liquid crystal phase (bright area) is forming from the liquid phase (dark area). (Courtesy of J. Goodby)

Plate 7. Polymer liquid crystal. This polyester sample is between crossed polarizers and in the nematic phase. (Courtesy of E. Samulski)

Plate 8. Cano wedge. This chiral nematic liquid crystal sample is in the space between a flat piece of glass and a curved lens, which is in turn between crossed polarizers. The circular lines are disclinations. (Courtesy of P. Crooker)

Plate 9. Flat panel computer monitor. This TFT-LCD monitor utilizes in-plane switching and hence has a very wide viewing angle. It also can be used as a television. (Courtesy of Samsung Electronics)

Plate 10. Small surface stabilized ferroelectric liquid crystal (SSFLC) display suitable for a projection television. (Courtesy of Displaytech, Inc.)

Plate 11. Switchable light panels. The middle panel is in the opaque state, while the two outside panels are in the clear state. (Courtesy of Taliq Corporation)

Plate 12. Electronic book. An entire book is stored in memory and is visible page by page on a paperlike display. (Courtesy of Eastwick Communications)

Plate 13. Welding mask with liquid crystal darkening. The window of this mask turns dark instantaneously when the welding source is turned on and returns to a transparent state when the welding source is turned off. This allows the welder to keep the mask on as the welding source is turned on and off. (Courtesy of Hornell, Inc.)

Plate 14. Diagnostic foot insoles. Color photographs of a patient's feet (upper left) and of diagnostic foot insoles from the same patient (lower right). Parts of the feet with different temperatures show up as different colors when these insoles are used. Such temperature profiles allow for the diagnosis of circulatory problems. (Courtesy of Flexi-Therm)

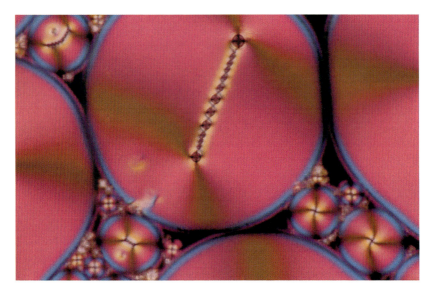

Plate 15. Nematic emulsion produced by mixing a nematic liquid crystal and water. The liquid crystal exists as large droplets (diameter about 50 micrometers) in water, but inside the liquid crystal droplets are smaller droplets of water (diameter about 1–5 micrometers). Due to the deformation of the orientational order in the liquid crystal, the small water droplets tend to attract each other and form chains. The sample is between crossed polarizers. (Reprinted with permission from P. Poulin, H. Stark, T. C. Lubensky, and D. A. Weitz, *Science* 275:1770. Copyright 1997 American Association for the Advancement of Science.)

Plate 16. Blue phase single crystals. Even though the blue phase is completely fluid, its red and green crystals sometimes grow with flat boundaries. In some cases facets are visible. (Courtesy of H. Stegemeyer and Th. Blümel)

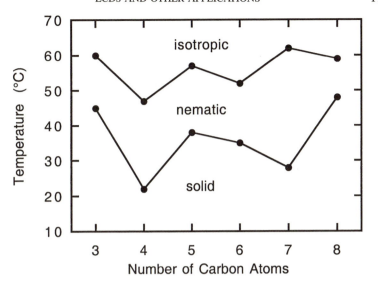

Fig. 7.5. The transition temperatures and phases for the third through eighth members of the Schiff base homologous series created by changing the length of the chain that does not contain oxygen. The compound with four carbon atoms is *p*-methoxybenzilidene-*p*-*n*-butylaniline (MBBA).

Fig. 7.6. Two examples of liquid-crystal-forming ester molecules. The NO_2 and CN groups are highly polarizable.

All the compounds described so far break apart at the central linkage group when exposed to moisture, high temperature, and ultraviolet light. The pure compounds therefore become a mixture of the original compounds and new impurities. These impurities usually do not form liquid crystalline phases, so both the transition temperatures and the liquid crystal temperature ranges change as the impurities build up. Displays engineered to work effectively at room temperature thus begin to work less effectively, and can even cease to work if the liquid crystal phase is shifted above or below room temperature. Although liquid crystal displays were manufactured using these materials, they only worked for about a year or two.

The breakthrough was achieved in the early 1970s by a group of chemists working under George W. Gray at the University of Hull in England. They eliminated the central linkage group completely by connecting the benzene rings together, forming biphenyl and terphenyl compounds. These compounds are very stable, thus eliminating one of the chief problems hindering the development of high-quality LCDs. A few examples of compounds based on the biphenyl or terphenyl central core are shown in figure 7.7.

One important feature of some liquid crystals is illustrated by some of the compounds shown in figure 7.7. Notice that some contain a short polarizable group on one end instead of a hydrocarbon chain. In many cases the presence of such a group enhances the liquid crystalline character of the molecule, even though it makes the molecule shorter. The reason for this apparent paradox is that the highly polarizable group causes the molecules to pair up slightly, resulting in a combined molecule (called a *dimer*) that is longer than the single molecule. This formation of pairs was briefly discussed in chapter 5, where it was pointed out that X-ray studies showed that in some cases the smectic layers have a width equal to one and a half molecular lengths (see figure 5.10). The most common polarizable groups used in developing liquid crystalline materials are the cyano (CN) and nitrous dioxide

Fig. 7.7. Three examples of biphenyl and terphenyl liquid crystal molecules, which are useful for display applications. The cyano (CN) group is the polarizable end group.

(NO$_2$) groups. The transition temperatures for some of these liquid crystals are also quite low.

STRUCTURE-PROPERTY RELATIONS

Over the years a huge number of liquid crystalline materials have been synthesized and their properties (transition temperatures, viscosities, electric susceptibilities, etc.) measured. As a result a great deal is known about how the molecular structure influences the material properties. We have already discussed a few aspects of this by giving some examples of homologous series and describing the quest for a room temperature, stable nematic liquid crystal. A fuller picture is warranted, although a complete description of how chemists synthesize liquid crystalline materials in order to optimize their properties for a specific application is beyond the scope of this chapter.

A schematic of the basic structure of a rodlike liquid crystal is shown in figure 7.8. Notice there are two core units, a bridging group between them, two groups that link the terminal units to the core groups, the terminal units themselves, and possibly lateral substituents on the core groups. The same basic structure will produce both the nematic and smectic phases, but a few generalizations are useful. Short terminal units and a central part of the molecule polarizable along the long molecular axis favor the formation of nematic phases. The choice of terminal units also is important, with an aromatic ring or polar group more likely to produce a nematic phase than a methyl group or hydrogen atom. In order for molecules to form a smectic phase, they must pack nicely into layers. Symmetrical molecules with two identical cores and two alkyl or alkoxy terminal units usually form a smectic phase. Anything that breaks the symmetry or makes it difficult for the molecules to pack tightly into layers serves to limit the possibility of smectic phase formation. Examples of this are the use of lateral substituents on the core (e.g., a fluorine atom or methyl group) or a central part that is relatively long compared to the total length of the molecule.

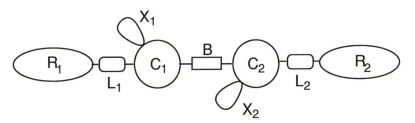

Fig. 7.8. Structure of a rodlike liquid crystal molecule. There are two core groups (C_1 and C_2), a bridging group (B), two terminal units (R_1 and R_2), two groups linking the terminal units to the cores (L_1 and L_2), and two lateral substituents on the cores (X_1 and X_2).

It was originally thought that highly polarizable core units such as benzene rings were necessary for liquid crystallinity. This is now known to be incorrect, as a good deal of work has been done with cyclohexane rings to produce liquid crystals with superior properties. The ability of compounds with cyclohexane rings to form liquid crystals probably stems from their ability to pack efficiently with a high degree of orientational order. A simple model that illustrates this point is one in which the cyclohexane unit is represented by a zigzag shape. Such a shape can pack quite well by fitting similar parts of the unit next to each other. Notice that such a packing also possesses a high degree of orientational ordering.

There are two basic ways to generate the tilted smectic C phase instead of the untilted smectic A phase. The first is to use alkoxy terminal units instead of alkyl terminal units. This is illustrated in figures 7.2 and 7.3, where the only difference between the two series is that an oxygen atom links the terminal chains to the core units in the series with a smectic C phase. Since the addition of the oxygen atoms tends to bend the terminal chains away from the long axis of the central part of the molecule, a simple zigzag picture is often used to explain the behavior. Zigzag molecules can pack tightly into layers, but only if parts of the molecules are tilted relative to a line perpendicular to the layers. The second method used to produce the smectic C phase is the addition of lateral substituents on the core units. A lateral fluorine atom is often used and is thought to produce a permanent dipole across the molecule that aids in the formation of tilted phases.

Many applications require that a liquid crystal have specific properties, so understanding how molecular structure influences these properties is extremely important. In general, the more rigid and less polarizable the core units, the higher the liquid crystal to isotropic transition temperature. Terminal units also affect transition temperatures, with the tendency for longer terminal chains to have higher transition temperatures. Figure 7.3 is an example of this, but figure 7.2 shows little effect as the terminal chains are increased, which is something that occurs in most series as the chains reach a certain length. Transition temperatures are also affected by short terminal units, with the larger and more weakly polar groups tending to have higher transition temperatures. Shorter and less polarizable central parts favor lower viscosity, probably due to weaker intermolecular forces. As expected, more highly polarizable central parts and the inclusion of permanent dipoles produce higher electric susceptibilities in liquid crystal phases. Incorporation of benzene rings in a molecule seems to be the best way to obtain materials with high birefringence. If a chiral nematic or chiral smectic C phase is desired, the molecule must contain a group that promotes molecular ordering with a slight twist between neighboring molecules. Such groups contain a *chiral center*, which is usually a carbon atom with four different substituents bonded to it. Usually the group containing a chiral center is added to one of the terminal units, but it has been incorporated into the central part of the molecule also. If the chiral center is in the terminal unit, there is a tendency for the pitch of the chiral liquid crystal phase to decrease the closer the chiral center is to the core unit.

SYNTHESIS OF LIQUID CRYSTALS

Liquid crystals are synthesized using the standard procedures of organic chemistry. By starting with available chemicals that resemble parts of the desired liquid crystal compound, chemical reactions are employed to link these parts together. For quite some time during the history of liquid crystal research, only very simple starting compounds were available. Many reactions were therefore necessary to obtain the final liquid crystal compound. More recently, complex starting compounds have become commercially available, so the synthesis proceeds with a fewer number of reactions.

Let us examine the synthesis of MBBA as an example of what actually must be done. MBBA can be formed in the reaction of *p-n*-methoxybenzaldehyde with *p-n*-butylaniline. Both of these compounds are shown in figure 7.9. The reaction produces MBBA (*p-n*-methoxybenzylidene-*p'-n*-butylaniline), which is also shown in figure 7.9, and water. This reaction is normally carried out in a mixture of solvents that provide the proper conditions for the reaction to take place, and the water produced during the reaction must be removed as the reaction proceeds. The MBBA is separated from the other compounds by filtering techniques and evaporation of the

Fig. 7.9. Starting compounds and the final compound in the synthesis of MBBA: (a) *p-n*-methoxybenzaldehyde, (b) *p-n*-butylaniline, and (c) MBBA.

solvents. The final step is to purify the MBBA, usually by recrystallization in appropriate solvents.

In many cases the two compounds that react to form the liquid crystal compound are not commercially available. The chemist must first perform reactions that form these compounds, starting with other materials that are available. For example, p-n-methoxybenzaldehyde is produced in the reaction of p-hydroxybenzaldehyde (a compound similar to figure 7.9[a] but with an HO group in place of the CH_3O group) with n-methylbromide (CH_3Br). In this reaction hydrogen bromide is also given off. Likewise, p-n-butylaniline can be produced in a series of three reactions starting with n-butylbenzene (a compound similar to figure 7.9[b] but with simply a hydrogen atom in place of the NH_2 group).

In general, these reactions are not simple ones to perform. If the conditions are not correct, the reaction does not proceed at all, or another reaction involving the starting material takes place, only to produce an undesired product. As if this is not enough, separation and purification of the products of these reactions are often not easy procedures. It takes a good organic chemist, skilled in these types of techniques, to efficiently synthesize liquid crystal compounds. Success is usually achieved after careful consideration of the details of the reaction together with some painstaking trial and error.

MIXTURES

With so many liquid-crystal-forming compounds available, it might seem that finding one suitable for any application would be quite easy. This is simply not true. Any one application demands that the liquid crystal have the proper temperature range, the proper response to an electric field, the required stability, the proper viscosity, and so on. For all the correct properties to occur in a single liquid crystal is improbable. Luckily, mixing two or more liquid crystals together allows the scientist or engineer to vary the properties of liquid crystals almost continuously. Many liquid crystals are miscible, with properties intermediate to the starting compounds. In many cases mixtures have wider temperature ranges for the liquid crystalline phase than either of the original liquid crystals. This is shown in figure 7.10 for a mixture of p-n-pentyl-p′-cyanobiphenyl (5CB) and p-n-octyloxy-p′-cyanobiphenyl (8OCB). The compound 5CB has a nematic range of 24°–35°C while 8OCB has a nematic range of 67°–89°C, neither of which is satisfactory for display purposes. However, a mixture of roughly 35% 5CB and 65% 8OCB has a nematic range of 5°–50°C, which is quite suitable for an LCD. A mixture for which the nematic range is as wide as possible due to a minimum in the solid-nematic transition temperature is called a *eutectic* mixture. The mixtures used in LCDs are eutectic mixtures, usually consisting of four to ten components. Two examples of fairly recent mixtures are E43, which is composed of biphenyls and terphenyls and has a nematic range from −10° to

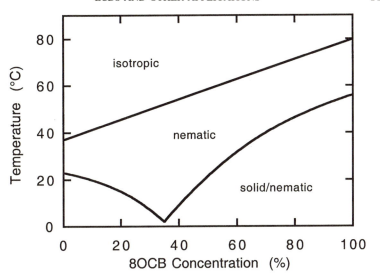

Fig. 7.10. Transition temperatures for mixtures of pentyl-cyanobiphenyl (5CB) and octyloxy-cyanobiphenyl (8OCB).

84°C, and NP1694, which is composed of phenylcyclohexanes and biphenyl-cyclohexanes and has a nematic range from −20° to 86°C. The commercial availability of stable liquid crystal mixtures with superior electrical and optical properties plus a wide nematic range has probably been the most important factor in the recent development of high-quality and reliable LCDs.

DYES

As discussed in chapter 6, dichroic dyes are used in some types of liquid crystal displays. To be useful, a dye must be fairly soluble in the liquid crystal and not degrade the characteristics necessary for LCD application. In addition, the dye must be stable and uncharged, must strongly absorb light, and must produce colors suitable for displays. In order for the dye to have very different absorption characteristics in the off and on states, the dye molecules must be well oriented along the director when dissolved in the liquid crystal. The dye molecules themselves, therefore, must be elongated in shape, and the liquid crystal itself must possess a high order parameter.

The two types of dyes that have found the most use in LCDs are those based on the azo and anthraquinone groups. As mentioned before, the azo group is chemically unstable, so these dyes degrade over time. This fact is unfortunate, because otherwise the characteristics of the azo dyes are superior to the anthraquinone dyes. Hopefully research going on right now will produce anthraquinone dyes or other dyes that are both stable and possess

(a)

(b)

Fig. 7.11. Two examples of dichroic dye molecules useful for liquid crystal display applications: (a) azo and (b) anthraquinone.

superior characteristics for display devices. Figure 7.11 shows two of these azo and anthraquinone dyes.

OTHER FACTORS IN LCD TECHNOLOGY

In all LCDs, the liquid crystal is contained between two pieces of glass. Although the choice of the type of glass used is not critical, research over the years has shown that using glass that contains a significant amount of sodium or other alkali ions causes some problems. These ions migrate to the surface of the glass and together with any moisture present form a charged layer that can change both the alignment and electric field pattern within the cell. Two techniques have been used to avoid this problem. One is to use borosilicate glass, which contains few alkali ions. The other is to coat the glass with a layer of silicon dioxide that blocks the alkali ions from coming into contact with any moisture. This problem can also be overcome by using plastic instead of glass. This is an important area of current research, since plastics offer the other advantages of lower breakability, high uniformity, and flexibility. The main problems to be overcome concern the facts that low-cost plastics are usually birefringent and some react chemically with the liquid crystal compounds.

A very thin layer (10–50 nm) of indium–tin oxide is usually used to make the transparent electrode pattern. These thin films transmit over 80% of the light incident on them and possess an electrical resistance low enough for proper operation of the cell. The transparent electrode material must form a precise pattern on both the front and rear glass surfaces. The two patterns are made to overlap only in the proper places. In this way the small strips of film that allow voltages to be applied to each element do not have an electrode across from them and therefore do not show up during operation of the cell. These electrode patterns are usually made by starting with glass fully coated with an indium–tin oxide film. A mask is then made through either a silk-screening or photolithography process and then applied to the fully coated glass. The unwanted indium–tin oxide is then chemically etched away. A technique that gives finer definition is to use glass that has an additional film of material resistant to etching on top of the indium–tin oxide. A mask is again made, but this time the mask determines what parts of the glass will be exposed to ultraviolet light. The resistive layer loses its resistance to etching after exposure to ultraviolet light, so the glass can be placed in contact with the etching solution without the mask in place. Actually, resistive films come in two varieties. In the second type the original film is resistive to etching only after exposure to ultraviolet light, so the mask must be the negative of the desired electrode pattern.

The most common alignment for LCDs is the homogeneous texture in which the molecules are parallel to the glass surface. This alignment is usually achieved by applying a thin coating of some long-chain polymer and then stroking the coating in a single direction with a soft material. This process may produce many small grooves in the polymer, all pointing in the same direction, or may simply stretch the polymer film in one direction. Either result forces the director to lie parallel to the direction of rubbing. Polyvinyl alcohol, some silanes, and polyimides are examples of coatings that work well. The first is susceptible to moisture, and the second produce such a thin layer that reliability is sometimes poor. The third are presently popular alignment agents. One other technique for creating homogeneous alignment is to evaporate silicon oxide onto the surface using a very oblique angle. This process does not produce the low tilt angles that can be achieved by using the rubbed films and is not as convenient in the manufacturing process, especially for large-scale displays. However, this process does achieve reliable results. Homeotropic alignment (where the liquid crystal molecules are aligned perpendicular to the glass surfaces) is usually achieved by coating the glass with an amphiphilic material. The polar end adheres to the glass while the nonpolar end (usually a hydrocarbon chain) points into the liquid crystal space. These oriented hydrocarbon chains at both glass surfaces serve to orient the liquid crystal material perpendicular to the glass.

The thickness of the LCD is a critical factor in the proper operation of the cell. Not only does the critical field depend on the thickness of the cell, but in certain types of cells an optical effect that is determined by the thickness of the cell is essential to cell operation. The thickness of the cell (usually 5–25 μm) must therefore be exactly controlled. The sealing material around the outside of the cell is usually used as the spacer. Typical materials are thermoplastics and thermosetting organic polymers. This technique sometimes does not achieve the required thickness uniformity, so an additional measure is to place glass fibers or beads of the proper diameter in the liquid crystal material. These dispersed spacers are invisible to the eye but ensure that the sealant material will cure with the cell at the proper thickness.

The polarizers used in many of the types of liquid crystal displays cause some of the most serious problems. For example, as a polarizer is made more efficient (polarizes the light better), it also transmits less light. The present compromise between brightness and contrast utilizes polarizers that transmit 40% to 45% of the light (remember that 50% is all that an ideal polarizer can transmit). These polarizers also lose some efficiency over time and can be easily scratched. At present, the polarizers are made from stretched polyvinyl alcohol films containing iodine that are placed between layers of cellulose acetate. If a colored polarizer is desired, either a colored filter is added or the polyvinyl alcohol in the polarizer itself contains a dye instead of iodine. The polarizer is attached to the glass surface with an acrylic adhesive and covered by a plastic layer for protection. A reflective polarizer is made by the addition of a metal foil reflector. It is an interesting fact that these polarizers and colored filters are an expensive part (materials and assembly) of LCD manufacture. Research into higher efficiency, lower cost, and longer-lived polarizers and colored filters is therefore important to the continued improvement in LCD technology.

Active Matrix Displays

One of the most exciting developments in LCD technology is the incorporation of semiconductor devices in liquid crystal displays. The idea is quite simple. If the liquid crystal cell itself does not have the optimum characteristics, simply combine it with an electrical device so that the combination does have the optimum characteristics. This would be easy if only one liquid crystal cell were involved, but LCDs contain many elements, each of which must be modified with the additional electrical device. This task is made easier, however, by the advanced semiconducting industry, which can fabricate large-scale integrated circuits made up of devices small enough to satisfy the requirements of an LCD.

This technique is best illustrated by considering a large LCD matrix display. As discussed in the last chapter, multiplexing simplifies making the connections to the individual pixels, but the threshold characteristics of the liquid crystal cell limit the number of pixels that can be multiplexed. A nonlinear device can be used to improve the threshold characteristics. A nonlinear device is simply any electronic device for which the output depends on the input in a nonlinear way. For example, if a nonlinear device is placed in series with the liquid crystal pixel as shown in figure 7.12, the pixel does not turn on until the voltage across the nonlinear device reaches its threshold. If the nonlinear device has a very sharp threshold characteristic, the liquid crystal cell plus nonlinear device also has a sharp threshold curve. In addition, the pixel turns on at a voltage equal to the sum of the threshold voltages for the nonlinear device and the liquid crystal cell. The increase in both the sharpness of the threshold characteristics and the threshold voltage itself increases the degree of multiplexing possible. Such a system is called *active matrix addressing* with two-terminal devices, since the nonlinear devices have only two terminals. Notice that each pixel still requires only two connections. As also shown in figure 7.12, the nonlinear device is fabricated on the glass with the electrode structure. Examples of nonlinear devices suitable for this purpose are back-to-back silicon diodes and metal-insulator-metal devices.

Even more interesting is the use of three-terminal instead of two-terminal nonlinear devices. The best example of a three-terminal device suitable for this purpose is a thin film transistor (TFT). Figure 7.13 shows the connections to such a device and how one might be fabricated on the glass substrate. The operation of an active-matrix-addressed display with three-terminal devices is very different. The three-terminal device acts as a switch; a small voltage

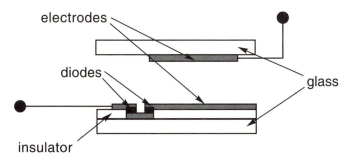

Fig. 7.12. One pixel of a back-to-back diode active matrix display. The liquid crystal is contained in the space between the two pieces of glass. The voltage is applied between the two electrodes shown with external connections, and therefore across both the back-to-back diodes and the liquid crystal cell in series.

Fig. 7.13. One pixel of a thin film transistor active matrix display. The liquid crystal is contained in the space between the two pieces of glass. The external connections show the three elements of the display that are connected to other circuit elements.

applied to its control terminal causes a larger voltage to be applied to the liquid crystal cell, thus turning it on. By using the proper three-terminal device, the liquid crystal pixel will stay on or off until the pixel is addressed again by the proper voltage to the control terminal of the three-terminal device. This effectively isolates the pixel from the voltages used to control other pixels. In addition, small variation of the voltage to the control terminal of the three-terminal device can cause precise variation of the brightness of the liquid crystal pixel. This type of display can therefore be engineered to produce an excellent gray scale.

The additional control available through the use of three-terminal devices obviously brings with it the additional complication of three connections to each pixel instead of two. Again, the highly developed technology of large-scale integration is capable of this much complexity. In fact, it is quite possible to drive these three-terminal active matrix displays directly, using a separate control line for each pixel. One idea that is currently being developed is to fabricate the electrical circuits necessary to send the proper voltages to these control lines on the glass substrate around the display itself. This would simplify making the connections between the display and other circuitry.

LIQUID CRYSTAL LIGHT VALVES

All liquid crystal displays are capable of varying the intensity of light emanating from different parts of the display. This feature has prompted researchers to investigate other possible applications besides displays. The general idea is that light of uniform intensity striking a liquid crystal panel can emerge from different parts of the panel with different intensities if the director configuration of the liquid crystal varies from place to place within the panel. The name for any device that can be used to vary the proportion of the incident light emitted is a *light valve*. If the light valve has the capability of varying the proportion of emitted light from place to place, the device is then known as a *spatial light modulator* (SLM). Liquid crystal light valves (LCLVs) and liquid crystal spatial light modulators (LCSLMs) have many possible applications, including light amplifiers, large-scale projection systems, optical data processing, and visible-to-infrared light converters.

The area of optical data processing is important and worthy of some discussion. The recent explosion of computer technology has demonstrated the power of fast computing. While further enhancements in the speed of electronic computers are expected, there are inherent limitations on how fast electronic devices can operate. Already there are efforts underway to increase the speed of computers by having them perform operations in parallel (simultaneous operations utilizing separate devices). This technique also has its limitations. It is clear that optical computing with increases of speed by a factor of at least one thousand has the potential to revolutionize computer technology. Not only is the actual speed of a typical optical device greater than current electronic devices; the ease of creating parallel optical channels further confirms the potential of this new area of development. The heart of such a technology is the SLM, which optically performs the same functions as the electronic switching devices in present-generation computers. The

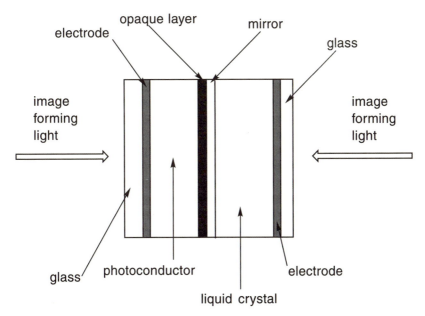

Fig. 7.14. A photoconductive image transfer device. The actual image is formed in the photoconductor by light coming from the left. The properties of the photoconductor cause an electric field pattern in the liquid crystal that is identical to the image. This image can then be seen when light strikes the liquid crystal from the right.

liquid crystal SLM is a promising candidate for such a system, and research into its use is now in progress.

Although most of the liquid crystal displays discussed in chapter 6 were electrically addressed (voltage signals controlled the display elements), we did see how a display could be optically addressed (scanning by a laser beam). In the case of liquid crystal light valves, however, much of the interest has been in optical addressing, since this method provides for finer detail (higher resolution) and is faster. For example, consider the LCLV shown in figure 7.14. The liquid crystal cell is placed in series with a photoconductive layer. The electrical resistance of such a layer varies inversely with the amount of light striking it. If a voltage is placed across both the liquid crystal cell and the photoconductive layer, a small voltage is applied to the liquid crystal where little light strikes the photoconductive layer, and a large voltage is applied to the liquid crystal where a great deal of light strikes the photoconductive layer. If the liquid crystal cell is designed to work in the reflecting mode, light incident from the image-forming side will produce a bright image on the reflective liquid crystal side, which corresponds to the

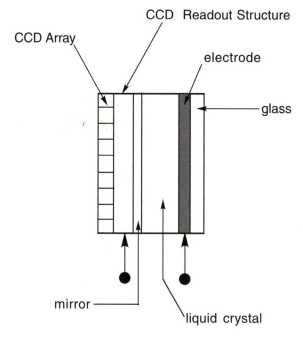

Fig. 7.15. A charge-coupled device (CCD) image transfer system. Light from the left forms an image in the CCD array that is transferred to the liquid crystal by the CCD readout structure. The liquid crystal image can then be seen when lit from the right.

pattern of light striking the other side. Notice that this is a *real-time* image, since there is no delay necessary in producing the image. Any system that must scan across an image in order to process it is much slower.

Liquid crystal spatial light modulators need not be used just to capture an image. Similar devices can be used to process an image already captured. This is part of a field called *optical processing*, and the advantage of LCSLMs is the speed at which they can perform various functions. As mentioned before, the liquid crystal devices in an optical processing system can be either electrically or optically addressed. For example, a liquid crystal light valve can be placed in series with an array of *charge-coupled devices* (CCDs) as shown in figure 7.15. Here a TV camera transfers the information it picks up in a scan to the CCD as voltage signals. The two-dimensional CCD array stores the information in the correct location as localized charge packets. Application of the proper control voltage to the CCD readout structure causes all of the charge to diffuse through to the liquid crystal cell, causing different amounts of deformation dependent upon the amount of charge present. In this way the entire two-dimensional image is transferred and read at once.

OTHER LCD-LIKE DEVICES

An interesting use of liquid crystal displays concerns the ability of some displays to resemble the printed page. Plate 12 shows an electronic book in which each page of the book appears on a black-on-white display. The entire book is stored in the memory of the device, and the reader simply presses a button to go from one page to the next. As shown in the picture, additional features such as highlighting and the ability to make handwritten notes are present. The number of books available in electronic form is increasing, and the ability of these displays to look exactly like a printed page is getting better. Producing a display with the characteristics of white paper turns out to be a difficult task. But once this is achieved, just think of the possibilities. How will we get our morning newspaper? Certainly it is not too much of a leap to envision downloading it electronically into a paperlike display device and then reading it by bringing different pages up on the screen.

An intriguing use of a liquid crystal light valve concerns the protective masks welders wear. Welding relies on creating a very small source of intense heat, capable of melting metal. A great deal of light is generated by these welding sources, enough to damage one's eyes. Yet welders must be able to see where they are working in spite of this bright light source. The conventional welder's mask has a window that allows only a very small portion of the light striking it to enter. With this darkened window, the working area is illuminated at a safe level, and as long as the welding source is operating, the welder can see just fine. The problem is that without the bright source of light, the darkened window allows so little light to enter that the welder can see virtually nothing. So welders must flip down the mask before turning on the source, perform the welding, and then flip up the mask after the source is no longer on. This is an inefficient and cumbersome procedure, especially when the welding task is complex.

Plate 13 shows a commercially available mask that uses a liquid crystal light valve to provide the darkening. Normally the window is transparent, and the welder can see everything in ordinary room light. But an electronic sensor in the front of the mask picks up the light from the welding source and causes a voltage to be applied to the liquid crystal light valve. Virtually instantaneously (at least fast enough so that no damage is done to the welder's eyes) the liquid crystal display darkens, and the welder can see the working area in the vicinity of the welding source. As soon as the source is turned off, the liquid crystal light valve becomes transparent, and the welder again can see with just the room lights on. Welders using this type of mask leave the window in front of their eyes, continually turning the welding source on and off without noticing much difference in what they can see.

These masks are quite sophisticated. In order to allow only 0.01% of the light to be transmitted when the welding source is on, two twisted nematic light valves in their dark states are used (one behind the other). There is also a filter to keep harmful ultraviolet and infrared radiation from being transmitted. The use of two light valves also produces an intermediate state when one is transparent and the other is dark (0.7% transmission). This is the condition when the mask has no power supplied to it. Thus the welder knows before turning the source on whether or not the mask is being supplied with power. Some models have a third twisted nematic light valve, the transmission properties of which can be controlled by the welder. This provides for a variable amount of darkening, which is important when different types of welding sources are used.

LIQUID CRYSTAL TEMPERATURE SENSORS

One of the most interesting (and beautiful) ways liquid crystals can be used is to measure temperature utilizing the selective reflection property of chiral nematic liquid crystals. As discussed in chapter 5, chiral nematic liquid crystals reflect light with a wavelength in the liquid crystal equal to the pitch. If white light, which contains a continuum of wavelengths, strikes a chiral nematic liquid crystal and propagates parallel to the twist axis, only light with a wavelength in the liquid crystal equal to the pitch is reflected back. To an observer, therefore, the liquid crystal appears to have the specific color as determined by the wavelength of light reflected. Since the pitch of chiral nematic liquid crystals changes with temperature, the color reflected by these materials also changes with temperature. In this way the temperature can be measured simply by observing the color of the chiral nematic liquid crystal. Such thermometers have been available for some time, being used to measure the temperature of a person's forehead, of the water in a fish tank, or of a room.

The usefulness of chiral nematic thermometers rests in the ease of designing devices that respond in the temperature range desired. Many chiral nematic compounds have a pitch that lies in the visible part of the spectrum, but the temperature at which this is true varies from one compound to the next. By mixing together the proper compounds, designers of liquid crystal thermometers can produce a material that reflects the different colors of the spectrum over whatever temperature range is desired.

In fact, very sensitive thermometers can be made in this way. Some chiral nematic liquid crystals possess a pitch that changes drastically over a very small temperature interval. Again, by mixing the proper liquid crystals, designers can produce a thermometer that reflects the visible spectrum over a temperature interval of less than one degree. Such thermometers are capable of revealing extremely small temperature differences and are useful in temperature-mapping applications. For example, a film of such a material placed in good thermal contact with the skin reveals in vivid color the temperature profile of the skin. This can be an important diagnostic tool for medicine, since many local medical problems (for example, a tumor located slightly below the surface) possess a temperature different from the surrounding tissue. If such an abnormality is present, it will show up on the liquid crystal film as an area of different color. The ability to test for slight temperature differences and possibly avoid surgery is sometimes important. Plate 14 is a picture using commercially available insoles to measure the temperature profiles of a person's feet. These allow for the diagnosis of circulatory problems using a very inexpensive and simple procedure.

An interesting example of where such films have found widespread use is in the "battery tester" that comes as a part of the package containing small batteries for portable equipment. This device is simply a metallic strip with a coating of temperature-sensitive liquid crystal on it. The metallic strip becomes wider from one end to the other (see figure 7.16), so its electrical resistance increases along its length (low at the wide end, high at the narrow end). When the strip is connected to the terminals of the battery, the varying resistance causes the temperature of the strip to vary continuously from being high at the narrow end to being low at the wide end. The color of the temperature-sensitive liquid crystal coating therefore changes first at the narrow end, and the color change gradually moves toward the wide end as the metallic strip grows warmer. If the battery is capable of producing the specified amount of current, the temperature at the low-resistive wide end will become high enough for the liquid crystal coating to change color. The

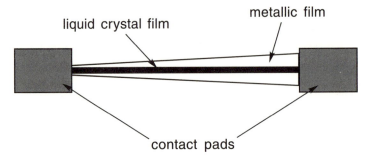

Fig. 7.16. "Battery tester" using a temperature-sensitive liquid crystal. The resistance of the metallic strip with liquid crystal coating is higher at the narrow end and lower at the wider end. When connected to a battery, the higher resistive end is warmer than the lower resistive end, meaning that the temperature-sensitive liquid crystal coating changes color at the narrow end first, with the change in color migrating toward the wider end. If the battery can supply the required amount of current, the color change reaches the wider end, and the user knows the battery is "good."

color thus migrates across the entire length of the metallic strip, and one knows the battery is working properly.

Liquid crystal temperature sensors have been used as a biofeedback mechanism in psychological therapy situations. All that is necessary is that the sensor be placed in close contact with the skin (between two fingers, for example). The advantage of the liquid crystal device is that it is much less expensive and easier to use than the more conventional monitors used for biofeedback (heart rate, blood pressure, EEG, etc.).

Such liquid crystal films can also be used for mapping the temperature profiles of almost any device. In some cases, such as precision ovens or electronic instrumentation, the information is important in order to find the proper design of the device. In other cases, the liquid crystal film can be used as a form of nondestructive testing. For example, poor electrical connections usually are slightly hotter than good connections. Although a device such as a printed circuit board may function properly, a poor electrical connection usually degrades over time to the point where the device no longer functions correctly. A chiral nematic film easily reveals a slightly warmer connection, thus allowing it to be fixed before the product is sold.

Some time ago, such liquid crystal materials were used in "mood rings." These rings reflected different colors depending on slight differences in temperature and were purported to indicate a person's different "moods." At least one manufacturer of children's toys has researched the idea of employing such liquid crystal sensors on figures or in games. The idea is that such figures could display "moods" or the strategy of a game might change during play depending on slight temperature changes. Some artists have also experimented with using chiral nematic liquid crystals as a medium. By using different mixtures, an artist can produce a work that displays different colors, but more important, the colors change with slight variations in the temperature of the room!

An important part of this technology is the ability to produce polymer films containing chiral nematic materials. Thin polymer films allow for good thermal contact between the temperature-sensing material and the object, and yet keep the liquid crystal from becoming impure through contact with other materials. These films are made through the process of microencapsulation discussed in chapter 6. This technique gives the designer good control over the purity of the encapsulated material, which is crucial for the design of sensitive thermometers. Chiral nematic liquid crystals can also be dispersed in a polymer film by mixing in the liquid crystal as the polymer sets (as is done in PDLC displays). Although the second technique is much easier, the purity of the encapsulated material cannot be controlled nearly as well, so this technique may have serious limitations for thermometry.

It is no mystery why many chiral nematic liquid crystals show drastic changes in pitch over a small temperature interval. Such changes usually occur in the chiral nematic phase in a temperature interval just above the transition to a smectic phase, and they reveal some basic physics about phase transitions in general. The smectic phase possesses layers that must buckle in order to twist, so it is much more difficult for smectic liquid crystals to twist than nematic liquids that do not possess layers. As the temperature decreases toward the chiral nematic to smectic phase transition, the molecules start to momentarily and only locally arrange themselves in layers. The temperature is too high for these layers to establish themselves permanently, but small groups of molecules begin to show some fleeting layer ordering. As the temperature gets closer and closer to the smectic transition, the length of time this local layering persists gets longer, and the size of the groups of molecules get larger. This makes it increasingly difficult for the chiral nematic liquid crystal to twist, so the pitch lengthens. Since all this can happen only near the smectic transition, the pitch change takes place over a tiny temperature interval just above the transition. At the transition, the entire material changes to a smectic liquid crystal, which usually has no twist at all (in other words, it has an infinite pitch). By mixing different materials together, the chiral nematic to smectic transition temperature can be made to occur at any temperature desired, so the temperature at which the chiral nematic reflects visible light is completely controllable. Proper selection of compounds in the mixture also allows the interval over which visible light is reflected to be varied in size. Figure 7.17 shows some data for a few mixtures prepared for commercial application. One type of thermometer uses dots containing different mixtures that respond over a temperature range of about two degrees, but at temperatures that differ by about two degrees. At any one temperature, therefore,

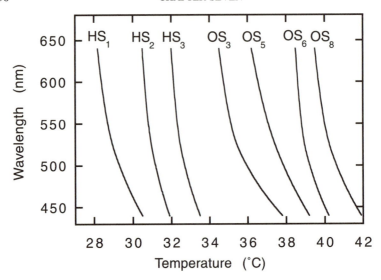

Fig. 7.17. Wavelength versus temperature curves for seven chiral nematic mixtures made for the purpose of application. Each mixture reflects all the colors of the visible spectrum over a two-degree temperature interval, but each mixture does this at a slightly different temperature.

only one dot shows any color. The temperature is read by reading the number written next to the colored dot.

OTHER APPLICATIONS

Some less important applications of liquid crystals rely on the general ability of liquid crystals to show their delicate response to various other changes in the environment. For example, chiral nematic liquid crystals change their pitch in response to pressure changes in much the same way they respond to temperature changes. Fairly large pressure changes are necessary to produce these pitch changes (many atmospheres), but a chiral nematic device kept at a constant temperature displays different colors for different pressures and therefore can be used as a pressure sensor. Likewise, exposure of chiral nematic liquid crystals to certain vapors causes a change in the pitch and a corresponding change in the color of light they reflect. Such devices can be used as sensors to detect low concentrations of these vapors. Ultrasonic waves striking an object cause its temperature to rise slightly. Therefore liquid crystal films can be used to detect the presence of these waves.

The Freedericksz transition discussed in chapter 3 is the basis for devices that can measure the strength of electric and magnetic fields. Since an electric or magnetic field changes the director configuration in a thin sample of liquid crystal, the properties of this thin sample are altered with a change in field strength. One can monitor an electrical, magnetic, or optical property of the sample (for example, its capacitance), and use it to measure the strength of the field. A device with a digital readout of the field strength is very feasible, although problems remain in making the device accurate over a long period of time.

Few liquid crystal applications utilize the mechanical properties of these phases, but at least one possibility is being explored. The viscosity of certain liquid crystal phases (for example, the smectic B phase) is much higher than the viscosity of the isotropic liquid phase. If such a material is used as the hydraulic element in brakes, clutches, or bearings, the amount of friction in these components depends on the phase of the material. In theory, temperature or pressure changes within the device could produce phase changes and thereby vary the characteristics of the device over quite a large range. The idea is to produce "smart" devices, which utilize the phase change to gain advantages not possible with conventional liquids. Our theoretical knowledge of exactly what happens when liquid crystals are used this way is lacking at this time, so a good deal of theoretical and experimental work will be required before these devices become a reality.

One last application of liquid crystals is a scientifically important one. In many instances, scientists and technicians perform measurements on substances that have been dissolved in some liquid. Two important examples come to mind. First of all, chemical reactions frequently are carried out in a liquid medium. Second, a sample containing various substances can often be divided into its constituent substances by forcing the mixture to pass through a liquid. The rates at which the different components of the mixture move through the liquid are different, so the various components come out of the end of the apparatus at different times. This area of analytical chemistry is called *chromatography*, and it can be used to investigate what is in an unknown mixture or to actually separate a mixture into its component parts. This technique is extremely important to many of the analyses done in chemistry, biology, and medicine.

Since the solvents used in both of these situations are liquids, the molecules of the substance to be investigated are situated in an isotropic environment. Sometimes using an anisotropic solvent (i.e., a liquid crystal) changes how the substance behaves in a way that is extremely important. For example, chemical reactions sometimes occur differently in a liquid crystal than in an isotropic liquid. This allows chemists to control the reaction in a way not possible otherwise. Likewise, the rates at which different molecules pass through a liquid crystal are sometimes not the same, whereas the rates through an isotropic liquid are the same. Molecules that differ from each other in very minor ways can often be separated using a liquid crystal medium when other methods fail. Liquid crystals are being used with increasing frequency as an anisotropic medium for such studies, and the power of this technique is only now starting to be exploited.

Although these last few applications have been presented as less important, one cannot be sure this situation will always be true. The ability of liquid crystals to perform one of these functions may become the cornerstone for a future technology of huge proportions. Perhaps someone writing ten years from now about the applications of liquid crystals will devote more attention to one of these possibilities than to LCDs. Time will tell, but the delicate response of liquid crystals to almost any change in their environment ensures that researchers will find them useful for a wealth of purposes. How important they become and in what areas they become most important are questions to be answered by scientists and engineers presently at work in laboratories all over the world.

Chapter 8

LYOTROPIC LIQUID CRYSTALS

IN OUR DISCUSSIONS so far, I have concentrated on liquid crystal systems in which temperature is the variable that determines which phase of matter exists. Two other possibilities have been pointed out. The first was pressure, and throughout my discussions I have assumed that the external pressure exerted on the material is equal to atmospheric pressure. If the pressure on a sample is allowed to vary considerably, phase changes result just as if the temperature had been changed. In fact, pressure works in the opposite direction as temperature. Raising the pressure while keeping the temperature constant causes many of the same phase transitions one would observe if the temperature were lowered with the pressure constant. The second possibility I discussed was concentration in a mixture of two liquid crystals. As evident from figure 7.10, phase transitions result when the concentration of one component of a mixture is changed while the temperature is held constant. In all these cases, however, temperature remains the most important variable in determining phase behavior—hence the name *thermotropic* liquid crystals for all these materials.

There is another class of substances that displays liquid crystalline behavior only when mixed with another material. Although temperature still is an important variable in determining the phase present, the concentration of one component with respect to the other is far more important. These materials are called *lyotropic* liquid crystals and are just as interesting and important as thermotropic liquid crystals. For example, the "goo" that sometimes collects in the bottom of your soap dish is a lyotropic liquid crystal phase of the soap/water mixture. Likewise, every one of the cell membranes in your body owes its structure to the liquid crystalline nature of the phospholipid/water mixture. Lyotropic liquid crystals are both scientifically interesting and technologically important, as will become evident in the following discussion.

AMPHIPHILIC MOLECULES

All of us have heard the statement that oil and water do not mix. Let us imagine ourselves as chemists, mixing various liquids together in hopes of gaining some insight into why certain liquids mix while others do not. The results of our investigation would show that oil and water are representative of two general classes of liquids. Liquids belonging to the same class usually mix, while those belonging to different classes do not mix. This statement is far too simplistic, because there are liquids that are not easily classified into one class, and other variables such as temperature turn out to be important. But as a general description of the mixing phenomenon, it forces us to investigate what is different about the molecules in each class that causes this behavior.

To answer this question, we might consider the molecular structure of molecules representative of each class. In doing so we would discover that the molecules belonging to the class containing water are *polar*. This simply means that the bonding together of atoms in these molecules results in an uneven distribution of electrical charge. One part of the molecule is slightly positively charged and another part is slightly negatively charged. On the other hand, the atoms of the molecules belonging to the class containing oil are bound together in a way that distributes the electrical charge fairly uniformly. These molecules are *nonpolar*. Figure 8.1 shows the molecular structure of water (with the charged parts indicated) and decane (a typical oil).

Interesting phenomena occur when a molecule is composed of two parts, one that by itself would mix with water and another that by itself would not. The part that would by itself be soluble in water is called the *hydrophilic* (water-loving) group, and the part that would by itself not be soluble in water is called the *hydrophobic* (water-fearing) group. These molecules can display characteristics of both classes, so we call them *amphiphilic* molecules (loving both kinds). The two most important types of amphiphilic molecules are soaps and phospholipids, both of which will be discussed in detail later. Amphiphilic compounds are also known as *surfactants*. As we shall see, the

(a)

(b)

Fig. 8.1. A polar molecule, water (a), and a nonpolar molecule, decane (b).

reason for this is that amphiphilic molecules tend to migrate to the surface of a liquid. This fact has important scientific and technological implications.

A typical *soap* molecule is shown in figure 8.2(a). It consists of a polar head group formed by the bonding between a sodium atom and a carboxyl group, and a nonpolar group consisting of a hydrocarbon chain. A common symbol for this molecule is also shown. It consists of a small circle to represent the hydrophilic head group and a zigzag line to represent the hydrophobic end chain. Soaps can be made using many other polar head groups together with hydrocarbon chains of various lengths. Some soaps have two hydrocarbon chains connected to the polar head group.

A typical *phospholipid* molecule is shown in figure 8.2(b). It consists of a larger polar head group containing a phosphorous atom and two nonpolar hydrocarbon end chains. Its simplified symbol is a small circle with two zigzag lines attached. Again, many different phospholipids exist due to the large number of possible head groups and hydrocarbon chains of different lengths.

(a)

(b)

Fig. 8.2. A typical soap molecule, sodium laurate (a), and a typical phospholipid molecule, dipalmitoylphosphatidylcholine (b).

Structures Formed by Amphiphilic Molecules

If a small amount of amphiphilic material is mixed with water, it is possible for the molecules to go into solution. As the concentration of amphiphilic material increases, however, one of two possible structures begins to form. If the amphiphilic molecules have a strongly polar head group relative to the nonpolar part of the molecule, the amphiphilic molecules begin to arrange themselves into spheres, with the polar head groups on the outside and the hydrocarbon end chains toward the center. This structure is called a *micelle* and is stable as long as the amount of amphiphilic material is above a certain concentration (called the *critical micelle concentration*). If the head group of the amphiphilic molecule is not strongly polar relative to the hydrophobic part, the molecules begin to form spherical *vesicles*, in which double layers of amphiphilic molecules (called *bilayers*) form a shell with water on the inside and outside. Cross sections of both of these structures are shown in figure 8.3.

It is not difficult to understand why both of these structures are stable. The hydrophilic head groups on the outside of the structures are in contact with the water molecules, while the hydrophobic end chains are in contact with each other and shielded from the water by the polar head groups. If the concentration of amphiphilic material is increased further, more micelles or vesicles form. In some cases the size and shape of the micelles or vesicles remain fairly constant as the number of them increases. In other cases the shape of the micelles changes from spherical to cylindrical. In some cases

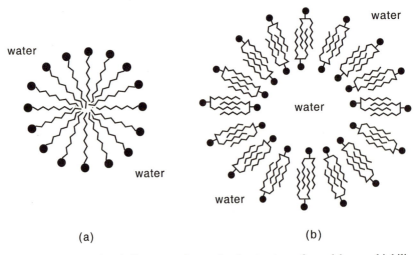

(a)   (b)

Fig. 8.3. Cross-sectional diagrams of two simple structures formed by amphiphilic molecules in water: (a) micelle and (b) vesicle.

vesicles with several bilayers form (each inside the next). The structure of these vesicles resembles an onion, but differs in that water occupies the region between each bilayer and the next.

It should be noted that similar structures begin to form if amphiphilic material is added to a nonpolar liquid such as oil. In this case the micelles or vesicles form with the polar head groups toward the inside and the nonpolar end chains toward the outside. Cross sections of these so-called *inverted structures* are shown in figure 8.4. These structures also can change in shape or bilayer number as the amount of amphiphilic material is increased. Although I will concentrate my discussion on the structures formed in polar solvents such as water, keep in mind that corresponding inverted structures are also possible.

If the concentration of amphiphilic material is increased even more (usually in the vicinity of 50%), a point is reached where the micelles or vesicles combine to form larger structures. One such structure is called the *hexagonal phase* (sometimes called the *middle soap* phase), in which long cylindrical rods of amphiphilic molecules arrange the long axes of the rods in a hexagonal array. Another common structure that forms at even higher concentration is the *lamellar phase* (sometimes called the *neat soap* phase), in which the amphiphilic molecules form flat bilayers separated from each other by water. One less common phase sometimes forms at concentrations between the hexagonal and lamellar phases. This *cubic phase* (sometimes called the *viscous isotropic* phase) is composed of spheres of amphiphilic molecules that arrange themselves in a cubic lattice. Whether the spheres are isolated

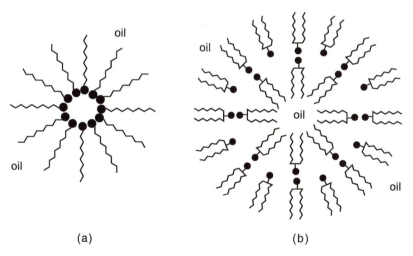

Fig. 8.4. Cross-sectional diagrams of two simple structures formed by amphiphilic molecules in oil: (a) micelle and (b) vesicle.

from one another or connected in some way is at present an open question. Cross sections of the hexagonal and lamellar phases are shown in figure 1.14. Undoubtedly, one of these structures is responsible for the "goo" in your soap dish.

These structures are liquid crystalline. The amphiphilic molecules diffuse throughout each of these structures, but in doing so orient themselves on average according to the structure. In some of these structures positional order is also present, so there are many similarities to smectic liquid crystals. Unlike smectic liquid crystals, however, the density of amphiphilic molecules varies drastically from point to point. The structures themselves are composed of mostly amphiphilic molecules, while the water between the structures contains a relatively small number of single amphiphilic molecules. The amphiphilic molecules are free to diffuse throughout both the structures and the water. They do so maintaining both the order of the structures and the large concentration difference between the amphiphile-rich structures and the water-rich solvent.

There has been a great deal of research into the identification of which polar head groups cause the formation of these structures and which do not. From this work we have learned that the head group need not be ionic (formed by the "donation" of an electron by one atom to another) in order for these structures to form. Likewise, not all polar, nonionic head groups are effective in producing these structures. The formation of these liquid crystalline phases is obviously the result of a delicate balance between the hydrophilic and hydrophobic tendencies of the amphiphilic molecule, and thus may turn out to be a difficult process to fully understand.

You should not be surprised to find out that temperature has a significant effect on the stability of these phases. In fact, none of these phases can form if the temperature is not high enough. The reason for this is that all these structures demand that the amphiphilic molecules be able to move relative to one another. If the temperature is too low, the molecules tend to form rigid crystalline structures. When mixed with water at this temperature, crystals of amphiphilic material simply exist in contact with the water. The temperature above which crystals do not form and these liquid crystalline structures do form is called the *Kraft temperature*. The Kraft temperature increases slightly as the concentration of amphiphilic molecules increases.

In order to summarize the behavior of these systems as both the concentration of amphiphilic material and the temperature is varied, we must resort to the same type of phase diagram used to describe how a mixture of two liquid crystals behaves. The concentration is plotted along the horizontal axis, with increasing amphiphile concentration to the right. The temperature is plotted along the vertical axis. At any point on the diagram one phase is stable, and curves divide regions where different phases are stable. The phase diagram for a typical soap is shown in figure 8.5. Note the curves denoting the critical micelle concentration (the nearly vertical dashed line) and the Kraft temperature (the curve separating the crystal and water part from the rest of the diagram). The phase diagram for a typical phospholipid is shown in figure 8.6. The critical concentration for the formation of vesicles is not shown on this diagram because it occurs at an extremely small concentration. Also, as explained in the next paragraph, the concept of the Kraft temperature is not nearly as useful for phospholipids. In both phase diagrams, additional liquid crystal phases (usually inverted phases) exist at very high concentrations, and various crystalline phases form at low temperatures.

The *gel* phases of certain phospholipids are quite interesting. The amphiphilic molecules form bilayers much like the lamellar phase, but the hydrocarbon end chains are rigid, much as they are in the solid phase. This is unlike the lamellar phase, in which the hydrocarbon end chains are free to adopt changing configurations, much as they are in the liquid phase. The rigid end chains of the gel phase allow the phospholipid molecules to arrange their polar head groups in a closely spaced hexagonal array.

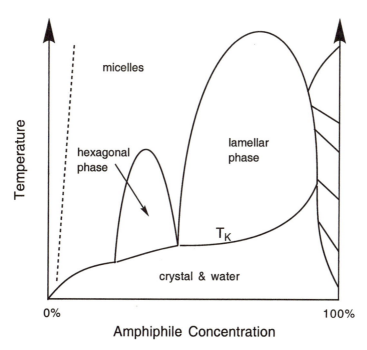

Fig. 8.5. Phase diagram for a typical soap. The nearly vertical dashed line shows the minimum concentration for micelle formation. The Kraft temperature, T_K, is the line separating the crystal/water part of the diagram from the rest of the diagram. Various liquid crystal phases occur in the region very close to the 100% concentration axis.

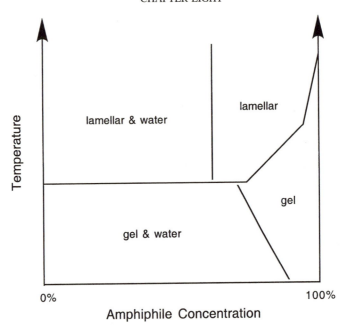

Fig. 8.6. Phase diagram for a typical phospholipid. Various liquid crystal and solid phases occur at high concentration and have been omitted from the figure.

Many substances have more than one gel phase, with the orientational and positional order increasing with each phase change as the temperature is lowered. Since no drastic change between liquid crystalline micelles and solid crystals mixed with water takes place, defining one of these transition temperatures as the Kraft temperature is problematic. Finally, it should be pointed out that the bilayers are rippled in some gel phases. This is not true of the lamellar phase.

THE MISCIBILITY GAP

Nothing underscores the delicate shift in balance between opposing tendencies that takes place in the formation of these structures better than the behavior of certain nonionic amphiphiles. Just like their ionic counterparts, these amphiphilic compounds first form micelles and then the various liquid crystal phases as the concentration of amphiphilic material increases. However, if the temperature is high enough, the sequence is quite different as the concentration of amphiphile increases. At very low concentrations micelles form as usual, but at a certain concentration they break up, and the system forms two separate phases, one rich in water and the other rich in amphiphile. If the concentration is further increased, the micelles again form, and the two

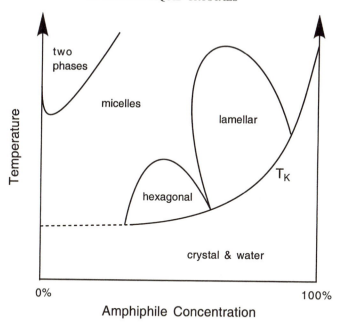

Fig. 8.7. Phase diagram for dodecyldimethylphosphide oxide in water. The region labeled two phases shows the miscibility gap of this system.

separate phases combine to form the usual homogeneous phase of water and micelles.

At these higher temperatures, therefore, a region exists where the two compounds cannot be mixed (that is, they are not *miscible*), and researchers refer to this as the *miscibility gap*. This gap can be seen in the phase diagram of the nonionic amphiphile/water system shown in figure 8.7. The miscibility gap is the region labeled "two phases" in the figure. Note that the miscibility gap widens as the temperature increases. In some cases even this feature reverses itself, with the miscibility gap becoming narrower and even disappearing if the temperature is raised sufficiently high. This bizarre behavior is direct testimony that the formation of micelles in amphiphilic/water systems is the result of a delicate balance. Changes in temperature and concentration are capable of drastically changing the balance condition.

Amphiphilic Compounds in Water/Oil Mixtures

The situation is even more interesting if oil is added to the amphiphile/water mixture. Although oil normally is not miscible with water, the amphiphile/water mixture has little difficulty in incorporating it into its structure. Let us concentrate on the micelle phase of the amphiphile/water mixture. As oil is added, it accumulates inside the micelles in contact with the *oliophilic* (oil-loving) hydrocarbon chains of the amphiphilic molecules. Because no separate oil-rich phase forms, the amphiphile/water mixture effectively "dissolves" the oil. This process is why soaps and detergents are effective as cleansers. We can describe this phenomenon in a slightly different way. Imagine placing water and oil together in a glass. Shake or stir as you might, the two do not mix and eventually separate into two phases. Now imagine adding some amphiphilic material to the water and oil. After a good shake the liquid in the glass appears slightly cloudy but homogeneous, and it remains that way. The amphiphilic molecules have migrated to all the boundaries between the two liquids (it is a surfactant) and have effectively "shielded" one from the other. So water and oil do mix if some amphiphile is added.

Let us consider this process in more detail. If we start with a mixture of amphiphile and water in the micelle phase, the addition of a small amount of oil causes the micelles to swell slightly as oil is incorporated into their centers. These swollen micelles are completely stable. As more and more oil is added, the micelles continue to swell and eventually reach the point where the amphiphilic molecules actually are the barrier between one phase (oil) and another phase (water). At this point the system is an *emulsion*, in that it must be described as a combination of two phases. Amphiphilic compounds are therefore *emulsifiers*. For oil concentrations less than that necessary to form an emulsion, we call the single phase of slightly swollen micelles a *microemulsion*.

To partially describe what phases result from the mixture of three components, we can use the same type of phase diagram as before. The only problem is we must select the ratio of oil to water and keep it fixed. We can then add amphiphilic material to the selected oil/water mixture and determine what phases are present at various temperatures and amphiphile concentrations. A representative diagram is shown in figure 8.8. Notice that in addition to lamellar and emulsion phases, certain concentrations cause the oil and water to separate into either two phases (oil and water, both containing amphiphile) or three phases (oil, water, and amphiphile).

Another way to partially describe a three-component mixture is to keep the temperature constant and vary the relative amounts of all three components. All possible ratios of the three components can be represented as points inside an equilateral triangle, where each vertex of the triangle represents 100% of one of the components. To determine the amount of component represented by a point inside the triangle, simply drop a line perpendicular to the side opposite from the vertex representing 100% of that compound. The length of this line is a measure of the concentration of that compound on a scale where the height of the triangle is equal to 100%. Figure 8.9 shows a representative phase diagram. Note the liquid crystalline, emulsion, and two phase regions.

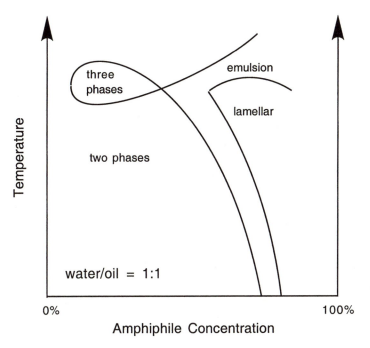

Fig. 8.8. Phase diagram for a typical tertiary system. The amount of water and oil is always equal, but the amount of amphiphile varies.

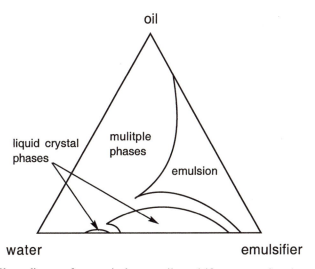

Fig. 8.9. Phase diagram for a typical water-oil-emulsifier system. See the text for an explanation of how to read a tertiary phase diagram.

APPLICATIONS

Our discussion of amphiphile/water/oil mixtures leads to the oldest and most important application of lyotropic liquid crystals, their use as detergents. Amphiphilic compounds have been used as soaps for over three thousand years, yet our knowledge of how they work is very recent. Many questions remain unanswered. For example, the phase diagrams such as the one in figure 8.8 show that certain amphiphiles are most effective in making oil soluble in water at a particular temperature. We do not completely understand why this is so. There can be no doubt, however, that this question touches an important aspect of technology. Just think of how useful it would be to develop detergents that work best at the temperature they encounter during a specific application. Perhaps a laundry detergent could be developed that really does wash in cold water as effectively as our modern detergents do at hot water temperatures. Needless to say, the detergent industry is huge, producing soaps, powders, creams, and foaming agents of all kinds.

A good deal of the research into amphiphile/water/oil systems is being done for the crude oil industry. A large portion of the oil trapped in porous rocks (up to 50%) remains there after an oil well runs out. If these oil fields could be flooded with a cheap amphiphile/water mixture that effectively incorporated the oil into its micelles, then the oil could be recovered by pumping out the amphiphile/water/oil mixture from the well. The amount of oil potentially available through such a process is huge.

The food industry is an active user of lyotropic liquid crystals as food emulsifiers. The addition of these compounds serves to maintain texture, color, flavor, or viscosity. Emulsifiers are used in mayonnaise, salad dressings, marshmallow, whipped cream, beer, cheese, ice cream, and jelly. One interesting example concerns the production of bread. Good mixing of the ingredients in bread is necessary in order for the bread to rise and bake properly. Some grains contain these liquid crystalline compounds naturally, so their action as emulsifiers guarantees a good mixture of the ingredients. Grains which lack these natural emulsifiers and normally produce inferior bread can be made to produce good bread by the addition of synthetic liquid crystals.

There may be some significant medical applications for lyotropic liquid crystal systems. For instance, some drugs are not very soluble in blood and consequently do not get carried by the bloodstream to all the important active sites of the body. Addition of a small amount of amphiphilic material to the blood could make the drug more "soluble" and therefore more efficiently transported throughout the body. The amphiphile would have to be nontoxic of course, but if it could be designed to work best at body temperature, only a small amount would need to be added to the bloodstream.

Another possible medical use is only slightly different. Drugs which cannot now be taken orally because the enzymes of the digestive tract break them down might be encapsulated by an amphiphilic compound that would protect the drug while in the digestive tract. The amphiphilic compound could be designed to dissolve in the bloodstream, thus releasing the drug.

THERMOTROPIC-LIKE LIQUID CRYSTALS FORMED BY MICELLES

Certain tertiary mixtures of a highly polar liquid, a slightly polar liquid, and an amphiphile form micelles or vesicles that are not spherical in shape. Usually this happens only if the concentration of amphiphilic compound is within a very narrow range. Within this range, rod-shaped micelles, disc-shaped micelles, or micelles with an elliptical cross-sectional area form. These micelles behave much like typical liquid crystal molecules, spontaneously orienting themselves with their axes along a preferred direction. Studies have uncovered examples in which the micelles orient to form just about all of the basic liquid crystal phases. The narrowness of this region indicates that this behavior is the result of a very delicate balance which sometimes causes nonspherical micelles to form. The phase diagram for one such mixture is shown in figure 8.10. Notice how small the range of concentration in the figure is.

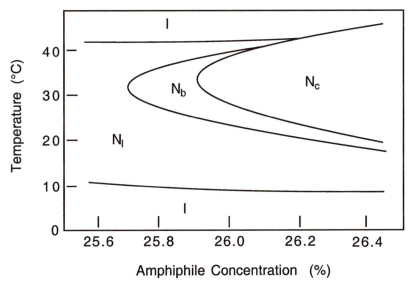

Fig. 8.10. Part of the phase diagram for the heavy water–decanol–cesium laurate system. The micelles are either rodlike (N_c), disclike (N_l), or more complicated in shape (N_b). Note that these phases are bounded at higher and lower temperature by the isotropic phase (I).

LIQUID CRYSTAL PHASES FORMED BY SELF-AGGREGATING SYSTEMS

Nature is full of examples of molecules that spontaneously aggregate in solution by forming weak bonds with one another. Rather than being amphiphilic in character, these molecules actually bond together weakly if the solvent conditions make this possible. Sometimes a slightly acidic solution is necessary for this to happen; in other cases there must be a certain amount of salt dissolved in the solution. Unlike many of the liquid crystal phases of amphiphilic molecules, the orientational and positional order of the molecules in these aggregates is probably not liquid crystalline in nature, but resembles more the type of order one finds in a solid. Some of the molecules that have this property are important biologically, a prime example being chlorophyll.

If the aggregates are anisotropic in shape and the concentration of the aggregates is high enough, liquid crystal ordering can develop among the aggregates. Perhaps the best-studied system is pseudoisocyanine (see figure 8.11). This molecule is referred to as a dye since it strongly absorbs light in one part of the visible spectrum. If the temperature is fixed, then as soon as the dye concentration reaches a certain value, the region of the visible spectrum in which the dye absorbs light shifts to longer wavelengths and narrows considerably. This dramatic change in the absorption region is a signature that the dye molecules are aggregating. At a slightly higher concentration the solution becomes birefringent, and the viscosity increases substantially. This is the onset of a liquid crystal phase in which the aggregates, which are probably rod-shaped, adopt long-range nematic-like order. It is interesting to note that the formation of the aggregates and the formation of the liquid crystal phase do not necessarily occur at the same concentration. The conditions under which the molecules form bonds with one another (mostly a chemical change) do not have to be the same conditions under which the anisotropic aggregates order themselves (mostly a physical change), although clearly the former change must occur for the second change to be possible.

The study of such systems is still in its infancy. While there has been a good deal of experimentation with self-aggregating systems, these studies

Fig. 8.11. Pseudoisocyanine dye. This molecule spontaneously aggregates under certain solvent conditions, and the aggregates can form a liquid crystal phase.

have usually taken place at very dilute concentrations where the chemistry of the systems can be investigated successfully. In fact, the study of these systems of *supramolecular assemblies* is gaining significant momentum, so there is no doubt that we will learn quite a bit more about the liquid crystal phases they form.

NEMATIC EMULSIONS

In what is somewhat the reverse of the polymer dispersed liquid crystal (PDLC) droplets described in chapter 6, interesting effects have been discovered when an immiscible isotropic liquid such as water is mixed with a nematic liquid crystal. Some of the water remains as spherical droplets in the liquid crystal, causing the director of the liquid crystal to adopt some surprising configurations. As shown in figure 8.12, in order to minimize the amount of director distortion introduced by the presence of the droplet, a defect in the director configuration spontaneously occurs near the water droplet. This defect represents a point at which the director is undefined, since the director points in different directions around the defect. We will

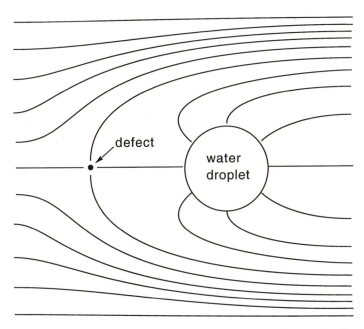

Fig. 8.12. Director configuration near a water droplet. In the figure, the director is parallel to the lines at all points. In order to minimize the amount of overall distortion, the director configuration spontaneously forms a defect near the droplet. These defect/droplet pairs tend to line up as shown in plate 15.

discuss such defects more completely in chapter 10. What is surprising is that the water droplet and defect pair creates a situation within the liquid crystal director configuration in which an attractive force occurs between each defect/droplet pair. Thus water droplets (with their associated defects) tend to form a chain within the liquid crystal, as clearly evident in plate 15.

While the observation of this attractive force was surprising, it did not take too long for researchers to understand the origin of the force. In fact, it is just another example of a liquid crystal adopting a director configuration that minimizes the amount of distortion. Creating a defect near each water droplet is one way the distortion is minimized. Having the defect/droplet pairs line up a certain distance from one another is a second way distortion is minimized. What is so provocative about this situation is that in some senses the defects and droplets are acting like electric charges and the liquid crystal director is acting like the electric field. The analogy between defects in liquid crystals and some physical fields can be quite close, especially when the mathematical equations governing the two situations are extremely similar.

ONE FINAL OBSERVATION

The amount of research performed on lyotropic liquid crystals over the last twenty-five years is small when compared to the amount done on thermotropic liquid crystals. However, there are indications that this is beginning to change. As we have seen, the potential applications of lyotropic liquid crystals lie in four important areas: detergents, food emulsifiers, oil recovery, and medical technology. The amount of money at stake for companies or individuals who successfully harness these applications is staggering. In the past, such a situation in other areas of science has always resulted in a dramatic increase in research activity. There is every reason to suspect that the same will be true for lyotropic liquid crystals.

Chapter 9

POLYMER LIQUID CRYSTALS

THE CHEMISTRY AND PHYSICS of polymers is a new and exciting area of science, which has already made a significant impact on our way of life. The future is even brighter; polymer science is an established and rapidly growing field of both basic science and engineering. As we shall see in this chapter, polymer liquid crystals have played and will continue to play an important role in these developments. Although it is beyond the scope of this book to include a full discussion of the entire field of polymers, we must begin with a general introduction to polymers, emphasizing those aspects important for understanding polymer liquid crystals. The different types of polymer liquid crystals and their properties are then discussed. The chapter ends with a description of a few areas where polymer liquid crystals have important applications.

In some sense, a more appropriate title for this chapter is "macromolecular liquid crystals." A *macromolecule* is simply a very large molecule, and as we shall see, polymers are a specific type of macromolecule. Since some of the discussion toward the end of the chapter concerns very large molecules that are not polymers in the traditional sense, the more general term (macromolecular liquid crystals) might seem to be a better choice. However, the advances in knowledge and the development of new technology in polymer liquid crystals over the last twenty years have been so important that it is indeed appropriate that they receive top billing.

POLYMERS

Any large molecule composed of a long sequence of repeating units is called a *polymer*. The unit that is the basic building block and that is repeated in the molecule is called the *monomer*. The monomer itself consists of atoms with chemical bonds between them; likewise, chemical bonds connect the monomers together in the polymer. The chemical reaction in which these bonds between monomers form is called the *polymerization* process. If we represent the monomer with the letter M, then we can represent the polymer by a series of such letters, MMMMMMMMMM. Figure 9.1 contains a typical example. The basic unit comes from the ethylene molecule, which is shown on the left. Part of the polymer (called polyethylene) is shown on the right, along with the symbol that we use to denote the polymer. The small

(a)   (b)

Fig. 9.1. Representation of a typical polymer. The ethylene monomer is shown in (a), and two representations (the full polymer and the repeating entity) of polyethylene are shown in (b).

letter n signifies that the unit between the brackets is repeated many times in the polymer.

The monomer can consist of just a few atoms as in the case of polyethylene, or its structure can be quite complex. The properties of the monomer ultimately determine the properties of the polymer, and we shall see that a huge range of physical properties is possible. Figure 9.2 contains some examples of polymers with which you are familiar. As you look at some of these examples, consider how different the properties of these materials are. The number of possible monomers is extremely large, which means that many different polymers can be produced.

The polymerization process rarely creates polymer molecules all of which have the same number of monomers. Therefore, any sample of the polymer material contains polymer molecules made from different numbers of monomers. To describe a polymer sample, we must state the average number of monomers in a polymer molecule (called the *degree of polymerization*) and state by how much the majority of the polymer molecules differ from this average number. For example, a sample of polymer might have a degree of polymerization equal to 100, with roughly two-thirds of the polymer molecules having a number of monomers differing from this average by less than 15 (that is, between 85 and 115). In general, it is quite difficult to force a polymerization reaction to produce polymers all with the same number of monomers, because the linking of monomers occurs quite randomly during the reaction. Thus there is little control over the length of each polymer molecule. The result is usually a statistical distribution in the number of monomers per polymer.

(a)

$$\left[CH_2-\underset{\underset{Cl}{|}}{CH} \right]_n$$

(b)

$$\left[CF_2 \right]_n$$

(c)

$$\left[CH_2-\underset{\underset{\underset{N}{\|||}}{C}}{CH} \right]_n$$

(d)

$$\left[\underset{\underset{O}{\|}}{C}-O-\bigcirc-\underset{\underset{CH_3}{|}}{\overset{\overset{CH_3}{|}}{C}}-\bigcirc-O \right]_n$$

Fig. 9.2. Four well-known polymers: (a) polyvinyl chloride (PVC), (b) Teflon, (c) Acrylan, and (d) Lexan.

Polymers can also be made from a chemical reaction in a mixture of two types of monomers. The result of this process is called a *copolymer*. If the two types of monomers (M and m) combine at random to form the polymer, a *random copolymer* results (MmMMMmMmmmMmMM). If the two monomers form short sequences of one type first (MMMM or mmmmm), which then combine to form the final polymer (MMMMmmmmMMMM-Mmmmm), a *block copolymer* results. Finally, if short sequences of one monomer (mmmmm) are attached as side chains to a very long sequence of the other monomer (MMMMMMMMM), a *graft copolymer* is formed. The properties of copolymers depend on many factors. These include the type of copolymer formed, the characteristics of the two monomers, the amount of one type of monomer versus the other, and the size of the short sequences in block and graft copolymers. Again, a huge number of possible monomers exist, so the combinations that can be chosen are even greater. In addition, the different ways the two monomers can be combined into a copolymer make the number of possible copolymers virtually limitless. The range of physical and chemical properties of all these copolymers is almost unimaginable.

As is true of all substances, polymers can exist in more than one phase. The solid phase of polymers can be either crystalline or amorphous. In the crystal phase, the polymer molecules occupy positions in a specific arrangement. In the amorphous phase, the polymers are arranged quite randomly but are not free to move to another position. The amorphous solid phase of polymers is also called the *glassy* phase, since glass possesses an amorphous structure. Often these two phases can coexist over a significant temperature range. A solid polymer, whether crystalline or glassy, will melt when heated. If it melts to a liquid crystal phase of some sort, it is called a polymer liquid crystal. In many cases the polymer melts to form a liquid without forming a liquid crystal phase. At high enough temperatures, a gas phase might be possible, but the polymer usually decomposes before this temperature is reached.

One additional comment should be made about these phase transitions in polymers. Because a typical polymer sample contains polymer molecules with varying numbers of monomers, the temperature at which a phase transition takes place is usually not as precisely defined as in materials consisting of identical molecules. The reason for this is not difficult to understand. The exact temperature at which a polymer changes phase depends on the number of monomers in it. If the sample contains molecules with different numbers of monomers, all molecules do not behave identically when the temperature is changed. The result is a phase transition that is spread over a range of temperatures as opposed to the phase transition in pure material, which takes place at one temperature. This phenomenon is quite general; mixtures of thermotropic liquid crystals show the same effect. How wide the temperature interval of the phase transition is depends on many factors, including the components of the mixture and the relative amounts of each component. The most important factor in the case of polymers is the degree to which the number of monomers in the polymer molecules varies within the sample.

THERMOTROPIC POLYMER LIQUID CRYSTALS

If polymers are formed from certain types of monomers, the polymer possesses liquid crystal phases in addition to the solid and liquid phases. Not surprisingly, these monomers closely resemble molecules that by themselves form liquid crystals, so they fall into two general classes. The first class consists of monomers that are fairly rigid, anisotropic, and highly polarizable. Just as in thermotropic liquid crystals, these monomers are either rod-shaped or disc-shaped. The second class of monomers causing liquid crystallinity in the polymer are amphiphilic monomers. Very little is known about the behavior of polymers formed from amphiphilic monomers, mainly because little work has been done in this area. Therefore, this chapter deals almost exclusively with polymer liquid crystals formed from rigid, anisotropic, non-amphiphilic monomers.

The monomers can be attached together to form a polymer in two different ways. If the monomers form a long single chain by attaching to one another, a *main chain polymer* results. If the monomers end up as branches extending away from the polymer chain, then a *side chain polymer* forms. In either case, the attachment to the polymer chain can be along either a short or long axis of the monomer. Diagrams of these four types of polymers are shown in figure 9.3, using both rods and discs to represent the monomers. How chemists get monomers to attach themselves in these various ways is an important question, which will be addressed when we discuss the synthesis of polymer liquid crystals.

The use of rigid and polarizable monomers sometimes makes things quite difficult for the polymer scientist. In some cases using such monomers produces polymers that melt at high temperatures, making the study of their liquid crystalline or isotropic phases quite difficult. In addition, polymers made from such monomers sometimes are fairly insoluble in the normal solvents. Again, this makes working with these polymers more difficult. In spite of these hardships, scientists have created many useful polymer liquid crystals and have uncovered general rules that help them create materials with the properties they desire.

For example, a great deal of experimentation has shown that useful polymer liquid crystals result when the monomer consists of a rigid section away from the point of attachment and a more flexible section near the point of attachment. The polymers shown in figure 9.3 are drawn using such monomers. The rod or disc represents the rigid part of the monomer, and the zigzag line represents the more flexible part. The flexible part near the point of attachment is called the *spacer*. Numerous studies have been carried out to investigate what effect the length of the spacer has in a polymer when

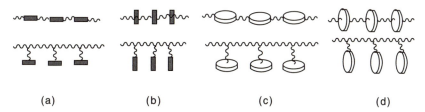

(a)   (b)   (c)   (d)

Fig. 9.3. Types of polymers that can be formed with anisotropic units (shown as rods or discs). The top row contains all main chain polymers and the bottom row all side chain polymers. The units can be attached along either their long or short axis.

the rigid, anisotropic section remains the same. Likewise, many other studies have kept the length of the spacer constant and have used various rigid groups to make different polymers. All this work has uncovered hundreds of polymer liquid crystals with nematic, chiral nematic, and smectic phases.

Whether main chain or side chain polymers, it is the rigid, anisotropic section of the monomer that displays orientational order in the liquid crystal phases. In the nematic phase, these rigid sections tend to point along a preferred direction just as in thermotropic liquid crystals. There is no positional order, and the other parts of the polymer display no orientational or positional order. Figure 1.13 shows the type of order present in nematic polymer liquid crystals for both main chain and side chain polymers. An elliptical shape has been used to represent the rigid, anisotropic part of the monomers. The same order parameter S, used to describe the amount of orientational order in liquid crystals, is also used for polymer liquid crystals. Its value describes how well the rigid, anisotropic parts of the polymer are ordered. Determinations of S for polymers reveal that the amount of order in polymer liquid crystals is close to that in liquid crystals. As can be seen by comparing figures 9.4 and 1.5, in some cases even the temperature dependence is very similar to the behavior in thermotropic liquid crystals.

In some cases, the preferred direction in a polymer liquid crystal is not constant but rotates in helical fashion just as it does in a chiral nematic liquid

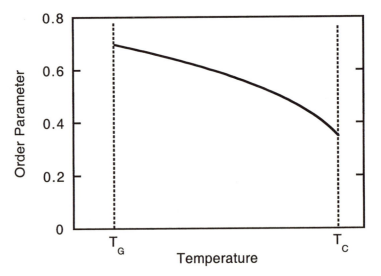

Fig. 9.4. Temperature dependence of the orientational order parameter in a typical polymer liquid crystal. T_C is the transistion temperature to the isotropic phase, and T_G is the transition temperature to the glassy phase.

crystal. As you might expect, monomers that resemble nematic liquid crystal molecules produce nematic phases in the polymer, while monomers that are similar to chiral nematic liquid crystal molecules form chiral nematic phases in the polymer. Finally, it is possible for the rigid, anisotropic sections of the monomer to tend to position themselves in layers as they orient in the liquid crystal phase. In this case a smectic liquid crystal phase is present in the polymer. Keep in mind in all these cases, that it is only the rigid, anisotropic part of the polymer that possesses order of any kind. The more flexible parts of the polymer, whether main chain or side chain, wander throughout the phase displaying little or no order of any kind. This fact is also evident in figure 1.13.

A polymer liquid crystal may possess one or more liquid crystal phases. Figure 9.5 contains an example of a polymer liquid crystal and the phases it forms. As is true for liquid crystals, many sequences of phases are possible. Because pure samples of these polymers form a liquid crystal phase, they are called *thermotropic* polymer liquid crystals.

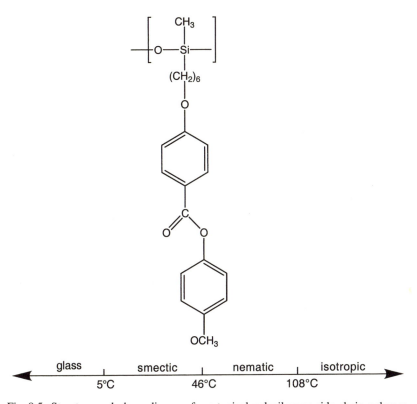

Fig. 9.5. Structure and phase diagram for a typical polysiloxane side chain polymer.

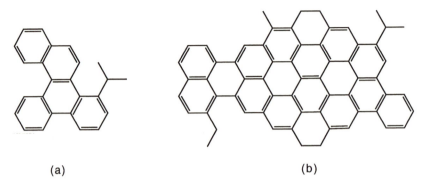

(a)   (b)

Fig. 9.6. Example of a network "monomer" (a) and "polymer" (b).

There is one other example of a thermotropic polymer liquid crystal that is quite different from the ones I have been discussing. Certain petroleum pitches and coal tars contain hydrocarbon molecules consisting of fused aromatic rings. When these materials are heated, the molecules polymerize into even larger molecules consisting of fused aromatic rings. Examples of both a smaller "monomer" and a larger "polymer" are shown in figure 9.6. Whereas we sometimes refer to the chainlike polymers as *linear polymers*, these large disclike molecules are called *network polymers*. The large disclike networks tend to orient with their planes parallel to one other, just as is the case for the molecules of a discotic liquid crystal.

Thermotropic polymer liquid crystals form when the polymeric material is heated to a point where the solid phase melts. For this reason, these liquid crystal phases are sometimes referred to as *polymer melts*. There is one other possibility. Some very large macromolecules (often polymers) form phases with orientational order when dissolved in ordinary solvents. In these cases, concentration is more important than temperature, so these phases are called either *lyotropic polymer liquid crystals* or *polymer solutions*.

Lyotropic Polymer Liquid Crystals

The first requirement for a macromolecule to form a liquid crystal phase in solution is that the molecule be fairly rigid. Rigid monomers that form strong bonds with each other represent one possibility. Macromolecules that adopt a single-or double-stranded helical structure with strong forces holding the structure in place also act as very rigid molecules and form liquid crystal phases. The second requirement is that the macromolecules must dissolve in the solvent. Liquid crystalline phases form only when the concentration is high enough so that these macromolecules constantly interact with one another. This cannot be achieved if the macromolecules do not dissolve in

the solvent to high enough concentrations. As previously mentioned, these
two requirements are often mutually exclusive. The more rigid a molecule is,
the less soluble it is. Scientists sometimes must use very strong solvents (like
sulfuric acid) to dissolve enough material. Still, many synthetic and natural
large molecules form liquid crystal phases in many solvents.

A good example of a class of synthetic polymer liquid crystals is known
as the *polyamides*. Three typical monomers from this class are shown in
figure 9.7. Notice that the monomer and the bonds between monomers are
similar to what one finds in the central, rigid part of some liquid crystal
molecules. In both cases the resulting molecule or polymer is fairly rigid.

Just about all of the large molecules that form liquid crystal phases due
to a helical structure are biologically important molecules. For example,
many derivatives of *cellulose* are polymer liquid crystals. Cellulose itself is
a polymer consisting of a linear chain of repeating units. The unit is derived
from *glucose*, a sugar, and is shown in figure 9.8. Notice that there are
chemical entities (H, OH, and CH_2OH) attached to the central part of the
monomer. If some of these are replaced with larger chemical entities, the

Fig. 9.7. Three example of aromatic polyamides with liquid crystal phases. The polymer shown in (c) is known as Kevlar and is used to produce high-strength fibers.

Fig. 9.8. The basic unit of cellulose, which is derived from the sugar glucose.

linear chain adopts a helical structure, gaining rigidity and producing liquid crystal phases.

Amino acids form the basic building block for the next type of polymer liquid crystal. As shown in figure 9.9, a typical amino acid possesses two quite different reactive groups. The carboxyl group (COOH) on the end of one amino acid can react with the amino group (NH_2) on the end of another amino acid to form a chain of amino acids. The resulting macromolecules are called *polypeptides*. Their structure is basically a long helical chain, which causes the macromolecules to be fairly rigid. When they are dissolved in water at high enough concentrations, liquid crystal phases occur.

Some polynucleic acids are also polymer liquid crystals. In these macromolecules, two chains form a double helix. The chains are composed of alternating sugar (like cellulose) and phosphate (like phospholipid) residues. The two helical chains are attached together by nitrogenous base pairs bound to the sugar residues in the chains. Deoxyribonucleic acid (DNA) is the

R = hydrogen or organic group

Fig. 9.9. The basic unit of polypeptide. Each unit is derived from an amino acid.

most important example. The double helical structure is very rigid, and these polynucleic acids form liquid crystal phases at high concentrations in water.

The last example of a macromolecule that forms a liquid crystal phase is a linear virus such as tobacco mosaic virus. Linear viruses consist of proteins (chains of linked amino acids) around a small amount of RNA (ribonucleic acid) or DNA. The structure is again a helix, and the virus is rigid enough to form liquid crystal phases.

Because all these macromolecules have helical structures, they interact in a fashion that prefers that the molecules not be exactly parallel to one another. As a result, some of these molecules form chiral nematic phases instead of nematic phases. Some even produce vivid colors due to selective reflection. One interesting feature, however, is that the pitch of these chiral nematic phases depends not only on the polymer but also on the solvent. In some solvents a certain polymer will form a right-handed chiral nematic phase and in another solvent a left-handed phase. In some cases the proper concentration in the proper solvent will produce a nematic phase. Obviously, the forces between these large molecules are due to both the helical molecules themselves and the interactions between these large molecules and the solvent molecules.

Synthesis of Polymer Liquid Crystals

If two different groups on the monomer can react to form chemical bonds between the two different groups on other similar molecules, then a method called *condensation* can be used to synthesize polymers. For example, consider a molecule with a carboxyl group (COOH) on both ends and another molecule with a hydroxyl group (OH) on both ends. These two molecules can bind together to form another molecule, releasing water during the reaction. This new molecule, with a carboxyl group at one end and a hydroxyl group at the other end, can react with the other similar molecules to form a polyester polymer. In a slightly different example, one molecule with carboxyl groups at both ends reacts with another molecule with amine end groups (NH_2), giving off water. The resulting molecule has a carboxyl group at one end and an amine group at the other end. These can polymerize with other similar molecules to form polyamides. Nylon can be synthesized in this way, and as we just discussed, amino acids can form polymers in a similar fashion. Both of these condensation examples are shown in figure 9.10.

It is possible to synthesize polymers without releasing a product such as water. This works effectively, for example, if the monomer originally contains an unreactive electron pair that can be made available for bonding. Consider the propylene monomer shown in figure 9.11(a). Reaction with a molecule with one unpaired electron (shown as R in the figure) can cause one of the electrons in the carbon-carbon double bond to pair with the unpaired electron

(a)

(b)

Fig. 9.10. Two examples of condensation reactions. In (a), terephthalic acid and ethy-lene glycol form the polyester Dacron; in (b), two monomers (r and s give the number of carbon atoms) form a polymer.

of the molecule. As evident from figure 9.11(b), the new molecule formed in this reaction still has one unpaired electron. This new molecule can react with other monomers with a double bond, just as the first molecule with one unpaired electron did. This is shown in figure 9.11(c). In this way, the polymer grows longer and longer as more and more monomers are added. In the case of propylene monomers, the result is the polymer polypropylene. This type of polymerization reaction is called *addition* and is very important in the production of well-known polymers such as polyvinyl chloride (PVC) and tetrafluoroethylene (Teflon).

One final example of a technique for synthesizing polymer liquid crys-tals is quite different. Instead of simply starting with monomers, one reacts monomers with a polymer. This method works well for the synthesis of side chain polymers; the starting materials are a long flexible polymer and a short rigid monomer. Quite simply, rigid monomers bind to the long flexible polymer to form a side chain polymer. A good example of such a *modifi-cation* reaction is the synthesis of the polysiloxanes. Figure 9.12(a) shows the repeating unit of a long flexible polymer, and figure 9.12(b) shows a rigid monomer with a reactive end group. The modified side chain polymer is shown in figure 9.12(c).

(a)

$$CH_2\!\!=\!\!CH$$
$$|$$
$$CH_3$$

(b)

$$R\!-\!CH_2\!-\!CH^{\ominus}$$
$$|$$
$$CH_3$$

(c) $$R\!-\!CH_2\!-\!CH\!-\!CH_2\!-\!CH\!-\!CH_2\!-\!CH^{\ominus}$$
$$\qquad\quad |\qquad\quad\; |\qquad\quad\; |$$
$$\qquad\quad CH_3\quad\;\; CH_3\quad\;\; CH_3$$

Fig. 9.11. An example of an addition reaction: (a) propylene monomer; (b) result of initial reaction with a free radical, R; and (c) result of two additional reactions in the formation of polypropylene.

(a)

(b)

(c)

Fig. 9.12. Example of a modification reaction: (a) monomer of a typical polysiloxane; (b) anisotropic molecule with reactive end group; and (c) monomer of the resulting polymer liquid crystal.

It is important to keep in mind that these polymerization reactions can proceed in a variety of ways. In some cases, the number of monomers in a typical polymer molecule increases as the reaction takes place, but in such a way that most of the polymer molecules have nearly the same degree of polymerization. The distribution of the number of monomers in a typical polymer molecule therefore stays fairly sharp during the reaction, but this distribution constantly shifts to higher and higher numbers of monomers per polymer molecule. In other cases, the polymer molecules with the highest degree of polymerization react with each other rather than monomer molecules, quickly producing a small number of very large polymer molecules. The distribution of the number of monomers per polymer molecule is quite different in this case. As soon as the reaction begins, there are some polymer molecules with a large number of monomers while the rest of the polymer molecules have just a few monomers. As the reaction proceeds, the number of very large polymer molecules increases while the number of very small polymer molecules decreases. There is no gradual shift of the distribution from lower to higher numbers of monomers, simply a gradual transfer from a majority of small polymer molecules to a majority of large polymer molecules.

How these polymerization reactions cease is also important. When the number of reactive end groups decreases to the point where there is little chance for further reactions between them, the polymerization process stops. Hopefully, this occurs because most of the reactive end groups have combined to form large polymer molecules. There are other possibilities, however. A reactive end group can bind to an impurity or to an unreacted starting group. Alternatively, a reactive end group can bind to a normally unreactive group (usually through the transfer of an electron). The point is that there are reactions that continue the production of longer and longer polymers, and there are reactions that terminate the production of such polymers. As the process proceeds, both types of reactions take place, until the number of reactive end groups is small. The result of a polymerization process therefore depends on the relative speeds of the two types of reactions during the process. With this in mind, it easy to understand why these polymerization processes normally result in polymers of very different chain length. Extreme care is required to synthesize polymers containing roughly the same number of monomer units.

STRUCTURE-PROPERTY RELATIONS

In main chain polymer liquid crystals, the distribution of chain lengths sometimes causes transitions to be spread out over wide temperature ranges. In such systems, two-phase regions are therefore not unusual. If the polymer is fairly rigid, usually due to short flexible units between the rigid units, it does not easily pack into layers, so nematic phases are more common than smectic phases. In fact, the flexible units between rigid units play a role similar to the terminal units in thermotropic liquid crystals. The longer the flexible units connecting the rigid units of the polymer, the higher the tendency to form smectic phases.

The spacer between the rigid units and the flexible main chain in side chain polymer liquid crystals acts in a similar fashion. The longer the spacer is, the more likely smectic phases are to form. Many of the structure-property relations discussed in chapter 7 for thermotropic liquid crystals carry over for the rigid units in side chain polymer liquid crystals. One important difference is that the phases observed in polymer liquid crystals tend to be more ordered than for the same units in thermotropic liquid crystals. For example, a rigid unit that shows no liquid crystal phase by itself may have a liquid crystal phase as a polymer liquid crystal. Additionally, a rigid unit that has only a nematic phase by itself may have a nematic and/or smectic phase when incorporated in a side chain polymer liquid crystal. Laterally connected side chain polymer liquid crystals, as illustrated in the lower diagram of figure 9.3(a), strongly tend to form nematic phases over smectic phases. Although the spacer units tend to isolate the properties of the main chain from the liquid-crystal-forming behavior of the rigid side chains, there is still a connection. The most important influence of the main chain is on the glass transition and liquid crystal to isotropic transition temperatures. The more flexible the main chain polymer, the lower the glass transition temperature, and to a much lesser extent, the lower the transition to the isotropic phase. Hence increased flexibility of the main chain tends to produce wider liquid crystal temperature ranges.

Side chain polymer liquid crystals with reactive groups can be made to form cross-links between the polymer chains and thus a three-dimensional network. Again, the rigid units still tend to adopt liquid crystalline order, but now the network structure prevents flow. These *liquid crystal elastomers* have mechanical properties very different from polymer liquid crystals, in that they return to their original shape and size after being deformed. In addition, some liquid crystal elastomers possess *piezoelectricity*, the property wherein an electric field is generated in the substance when it is deformed. Unlike solid piezoelectric substances such as quartz, the liquid crystal elastomer must deform one million times more in order to generate the same electric field.

APPLICATIONS OF POLYMER LIQUID CRYSTALS

Before we can appreciate one of the most important uses of polymer liquid crystals, we must first recognize why polymers in general are such useful materials. If a polymer in the liquid state is allowed to cool into the crystalline or glassy state, it possesses properties similar to the solid phases of other large organic molecules. Generally these properties are not all that useful for applications. However, if the molten polymer solidifies while it is flowing or when it has just stopped flowing, the long and normally tangled polymer molecules are stretched out and oriented in the solid phase. The properties of this material are very different and extremely useful for many varied applications. In other words, the usefulness of polymers depends on our ability to orient the long polymer chains. Since the polymer chains in the liquid crystal phases of certain materials are already ordered, these materials can be solidified with even greater ordering of the polymer chains. This fact leads to one of the most important applications of polymer liquid crystals: ultra-high-strength fibers.

All processes that utilize flow to orient polymer molecules involve the same three steps. First, the material must be put into its liquid or liquid crystalline state. This can be done by either heating the pure polymer or adding solvent to the pure polymer. Second, this fluid must be made to flow in a way that causes the polymer chains both to orient and to produce the proper final shape. Third, the fluid must solidify, either by cooling the pure polymer or by removing solvent from the polymer solution.

The second step is the most critical of the three. A typical way of heating a polymer and getting it to flow is to use a *screw extruder*. As shown in figure 9.13, polymer in its solid form (usually pellets) is introduced into one end of the screw extruder and pushed along by the action of the screw. Heaters along the outside of the extruder heat the polymer as it moves, causing it to melt and mix thoroughly. At the other end is a *die*, which forces the molten polymer through narrow channels. The narrowness of the openings causes the pressure inside the extruder to rise as the screw turns. This combination of high temperature, high pressure, and the geometry of the die is what causes the polymer molecules to orient. The highly ordered

Fig. 9.13. Diagram of a simple screw extruder. The solid is introduced through the funnel, heated and mixed as it is moved along by the screw, and forced through the die.

fluid polymer exits the die (usually in the form of a thin sheet or fiber) and cools quickly into a highly ordered solid.

There are several variants to this procedure. In a process called *blow extrusion*, a jet of air directed against the film exiting from the die causes it to form a long cylindrical bubble before it solidifies. The polymer material can be made quite thin by this process. All our plastic bags are made in this way. *Blow molding* is slightly different. A cylindrical piece of polymer is extruded around a thin tube that sits inside a mold. Compressed air is blown out of the thin tube, expanding the material until it takes the shape of the mold. Plastic bottles are made this way. If the fluid polymer is forced into a cavity of a certain shape to solidify, a solid object results. This process is called *injection molding*. Thin sheets of polymer can be fabricated by forcing the material to flow between two rollers. This *calendering* process can be used with bulk polymer to make thin sheets or with already extruded sheets to make them thinner. Finally, a process called *fiber spinning* produces threads made from twisted polymer fibers. Small orifices on a *spinneret* form multiple fibers that are twisted as they are drawn thinner and collected.

The ability of flow to orient polymer molecules results from the fact that the forces between the molecules are such that the molecules tend to align. In the absence of flow, the random motion of the molecules due to temperature overcomes these forces in the case of a liquid polymer and allows only a small amount of order in the case of a polymer liquid crystal. Flow induces additional orientational order that can overcome the random thermal motion in liquid polymers or increase the degree of orientational order in polymer liquid crystals. Of course, the conditions must be right for flow to have the desired effect. Before these processes can be used, therefore, much experimentation with different temperatures, polymers, flow speeds, die shapes, and so forth must be done.

Polymer liquid crystals can achieve extremely high orientational order, which is useful for certain applications. The best example is Kevlar, which is a fiber formed from the liquid crystal phase of a polyamide. Although only slightly denser than nylon (which is formed from the liquid phase), Kevlar is roughly thirty times stronger. In fact, pound for pound (steel is five times denser), Kevlar is stronger than steel. No wonder it has found widespread use. Automobile tires, bullet-proof vests, and mooring cables are only a few examples of Kevlar's practical applications.

Polymer liquid crystals respond to electric and magnetic fields, which means they can be used for display purposes. An electric or magnetic field can orient a polymer liquid crystal, thereby changing its optical characteristics. If the polymer liquid crystal is cooled into the solid phase, it is possible to freeze it in this order. Since in this case the order remains even after the field has been removed, a display using a polymer liquid crystal can have storage capability. For example, a laser can be used to produce local heating and raise the temperature into the liquid crystal phase. If no electric field is applied to the cell, the polymer cools in an unoriented state where the laser beam has been applied and thereby creates a permanent image. The image can be erased by applying the laser beam again but allowing the polymer to cool as an electric field is applied to the cell. Polymer liquid crystals have some properties that make them very attractive for display purposes. Liquid crystal polymers are relatively inexpensive, stable, and most important, can be easily fabricated into thin films.

As mentioned before, liquid crystals are a useful medium in which to perform a separation of substances (chromatography). Liquid crystal polymers offer one advantage for such purposes, in that they release fewer molecules as gas when they are heated to high temperatures. Under these circumstances, liquid crystal polymers are much more efficient than other materials in separating substances.

Ultra-high-strength materials and electro-optic displays are two very important areas of modern technology. Clearly polymer liquid crystals are destined to play even a larger role in these two areas in the future. However, there may be other significant applications around the corner. For instance, liquid crystal ordering of biologically important macromolecules is critical to the workings of cells. Is the function of certain structures within the cell, perhaps of RNA or DNA itself, affected by whether or not the macromolecules possess orientational order? The answer to this question seems to be yes, so knowledge of the processes involved might open the door to significant medical breakthroughs.

Chapter 10

THEORIES, DEFECTS, AND FLUID LATTICES

PREVIOUS CHAPTERS have certainly demonstrated that liquid crystals have been used for important technical applications over the last thirty years. What has not been discussed is that the study of liquid crystals has been just as important scientifically. As scientists have struggled to understand why certain molecules form liquid crystal phases and have certain properties, these scientists have found themselves asking some of the same questions as scientists studying other phases of matter. In fact, some important new ideas applicable to many phases of matter have been first shown to be valid in liquid crystal systems. In this chapter, I will first discuss some of the theories of both the liquid crystal phase itself and the transitions that take place between liquid crystal phases. I will then examine the defects that occur in liquid crystals and the reasons why they are important to our understanding of the liquid crystal phase. Finally, we will see how nature orders these defects to produce a truly unique phase of matter. In short, this is a chapter concerning the science of liquid crystals. Like all areas of science, the story of liquid crystal science is full of creative new ideas, brilliant experiments, and exciting surprises.

THEORIES OF THE LIQUID CRYSTAL PHASE

An observer of nature recognizes quickly that just about every process is quite complex. Yet the success of science rests in the ability of the human mind to uncover fairly simple ideas that allow us to see through this complexity and bring some order to diverse natural processes. In this sense, the achievements of science stand as a monument to the creativity, ingenuity, and perseverance of the human spirit. The attempt to understand the liquid crystal phase is an excellent example of this struggle to understand complex phenomena, because the liquid crystal phase results from the mutual interaction of enormous numbers of molecules. Even with our fastest computers, exact calculations involving such large numbers of molecules with all their simultaneous interactions are not possible. Therefore, scientists must dream up ways to capture the behavior of these large numbers of molecules without considering each one of them. What follows is a description of a few of these attempts.

Perhaps the most influential theoretical attempt to describe the liquid crystal phase is due to two German scientists, W. Maier and A. Saupe. Although their theory was formulated in 1960, it continues to be the starting point for a large amount of theoretical work. The *Maier-Saupe theory*, as it is called, begins with the assumption that the most important force between liquid crystal molecules is the *dispersion* force. The dispersion force occurs between two molecules that possess no permanent electric dipoles but that can possess induced electric dipoles. If the electronic structure of one molecule fluctuates (due to temperature effects) and gains a momentary electric dipole, the electric field of this dipole polarizes the other molecule slightly, thereby creating an induced electric dipole on the second molecule. This induced dipole on the second molecule produces an electric field that tends to strengthen the original dipole on the first molecule. The result of this interaction is that the initial spontaneous fluctuation results in two induced dipoles, which tend to (1) attract each other and (2) keep the two molecules aligned. Because the origin of this force is in the random fluctuations occurring in the electronic structure of molecules, the dispersion force is not exceedingly strong. For example, Maier and Saupe performed calculations that indicated that the dispersion force is proportional to the inverse sixth power of the separation distance. Compared to the force between two charges, which varies as the inverse second power, this decrease with separation is very rapid.

Even armed with the dispersion force between molecules, the Maier-Saupe theory must still deal with the problem of calculating what happens if millions of molecules mutually attract one another with this force. To overcome this problem, the theory takes the giant step of conjecturing that a single molecule in a sea of other molecules must experience a force that on average is the same for all molecules. That is, with so many molecules diffusing around randomly, the average effect on any one molecule must be the same for all molecules. The theory thus deals with the force a single molecule experiences in the sea of other molecules, and since it is the same for all molecules, one calculation will suffice. The force a single molecule experiences is obtained by combining conjecture with the fact that the basic force between molecules is the dispersion force. The conjecture is that a molecule will experience a strong force to point along the preferred direction if the molecules around it are highly oriented along this direction. If so, the force on a single molecule should be proportional to the order parameter S. This results in a force on a single molecule that represents an average or mean due to all the other molecules. For this reason the Maier-Saupe theory belongs to a class of theories called *mean field theories*.

To calculate how the order parameter changes with temperature, the Maier-Saupe theory calculates the thermodynamic average of $(3\cos^2\theta - 1)/2$ using the force on one molecule due to all the others. The average cannot be directly computed, because this force contains the order parameter S, which is the unknown the theory is attempting to calculate. The average of $(3\cos^2\theta - 1)/2$ is just the order parameter S, so the equation for this average is really an equation that says S equals some expression that also contains S (along with temperature, volume, etc.) This equation can be solved for S in terms of temperature and the other parameters. In this way, the Maier-Saupe theory produces a prediction for how the order parameter should vary with temperature in the nematic phase. Because all the details of the molecules are lost when working with this simple average force on a single molecule, the Maier-Saupe theory produces a single prediction for all liquid crystals. For some liquid crystals the Maier-Saupe prediction corresponds well with measurements of the order parameter. For some liquid crystals, however, the agreement is not very good. An example of order parameter measurements in two liquid crystals together with a graph showing the prediction of the Maier Saupe theory are shown in figure 10.1.

There have been many embellishments to the Maier-Saupe theory since 1960. For example, the average force experienced by a single molecule due to the other molecules has been derived using more forces than just the dispersion force. In another example, the fact that the molecules are not perfect rods but more like flattened rods has been taken into account in deriving the average force on a single molecule. The predictions of these theories depend on the relative strength of the various forces or the relative dimensions of the

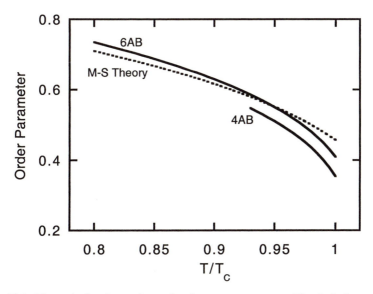

Fig. 10.1. Theoretical and experimental order parameter curves. The dashed curve is the prediction of the Maier-Saupe theory. The two solid curves show the results of order parameter measurements using two members of the PAA series. The horizontal axis is the temperature as a fraction of the nematic to isotropic transition temperature, which is different for the two compounds.

molecules. By making the proper selection for these parameters in the theory, good agreement between theory and just about all order parameter measurements has been obtained. The only problem is that it is difficult to select these parameters only from knowledge of the molecule being investigated and thereby predict how the order parameter should vary for a particular liquid crystal system.

Another theoretical approach that yields successful results emphasizes repulsive forces between molecules rather than attractive forces. When two molecules get very close together, the electrons of one molecule are close to the electrons of the other molecule, and the repulsive forces between these electrons dominate any other forces between the two molecules. Since these forces are relatively weak when the separation is not very small but increase sharply at some very small separation, these forces in effect keep the two molecules from occupying the same space. For this reason, these forces are called *steric* forces. Theories starting with steric forces seek to describe a system of rods that cannot penetrate each other. Such systems tend to align the long axis of the molecules, because the more they are aligned, the less chance there is for two molecules to collide.

The problem of performing the calculation for a huge number of molecules must also be overcome in this approach. Instead of performing averages over all molecules to calculate, for instance, the energy of the system, some theories consider only pairs of molecules. Other theories examine the situation in cases where the average is not difficult to perform, that is, for molecules much smaller or much larger than all the other molecules. An intermediate result that gives the correct behavior in the case of both large and small molecules is then selected for use in the rest of the calculation. One important aspect of these theories stems from the fact that the forces depend only on the shape of the molecules and the average distance between molecules. Temperature does not come into the calculation at all. Therefore these theories are not appropriate in describing thermotropic liquid crystals but work well in situations where temperature is not so important. For example, these theories give results that compare favorably with measurements made on a solution of rod-shaped macromolecules in a solvent. In this case the concentration of macromolecules is the important quantity, which can easily be taken into account by the theory. Both theory and experiment agree that at high enough concentrations of macromolecules, the molecules prefer an aligned liquid crystal phase over an unaligned isotropic phase.

One related type of theory is the *lattice model*. In these theories, the space the liquid crystal occupies is considered to be full of lattice points. A "molecule" therefore occupies a line of lattice points. The theoretical scientist considers all the different ways liquid crystal "molecules" can be placed in this space with each occupying a line of lattice points, but with no two "molecules" penetrating each other. An example of such "molecules"

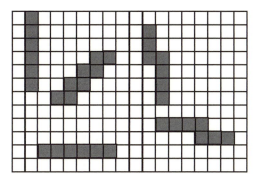

Fig. 10.2. Several examples of how molecules with various orientations can be represented in a lattice model.

in a two-dimensional lattice is shown in figure 10.2. Averages over all the different possibilities are performed (normally requiring a computer), followed by the calculation of various properties of the system. Since only steric forces are contained in this theory, the results are most appropriate to lyotropic liquid crystalline systems. It should also be pointed out that the size of the lattice cannot be very large before the computer calculation begins to take too long a time. A cubic space with roughly hundred lattice points on a side is about the present limit.

Another type of theory utilizes the *Monte Carlo* method. This is a method that computes the averages by artificially generating all the possibilities, but in a way that maintains consistency between the distributions of all the relevant parameters. How this is done is quite clever, so let us illustrate the method with a simple example. Suppose you want to find the value of π (3.1416) using the Monte Carlo method. This can be done by considering the square and the unit circle shown in figure 10.3. The area of the square is 4 and the area of the unit circle is π (area equals π times the radius squared). If you generate dots that are spread out throughout the square uniformly, the ratio of the areas ($\pi/4$) will give the fraction of the dots that fall within the unit circle. So on a computer you generate pairs of numbers (each randomly between–1 and 1) and call the first the x-coordinate and the second the y-coordinate. Eventually these numbers will fill the unit square uniformly. For each pair, you check if the point falls within the unit circle (is $x^2 + y^2$ less than 1?), and after many trials compute the fraction of points that fall within the circle. If you multiply this number by 4, you'll get a value for π. How good this value is depends on (1) the number of trials you perform, and (2) the quality of your generator of random numbers. Note that the points in the square are generated without considering anything about the situation, but the probability of a point falling inside the circle is the key in using the generated numbers to get the desired quantity.

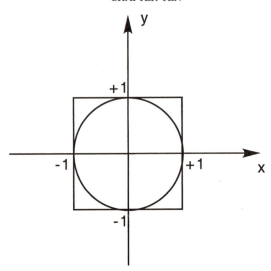

Fig. 10.3. Square with a side of two units and circle with a radius of one unit. The value of π can be found from these two figures using the Monte Carlo method.

In liquid crystal theory, a point inside the square is replaced by a possible configuration of all the molecules in a hypothetical liquid crystal using a lattice model. Such a possible configuration is generated without any regard for whether it is likely or not. The value for the quantity being examined (for example, the order parameter) is calculated for this configuration, but it is sometimes included in the average for all configurations and sometimes not, depending on whether it "wins a bet" against a generator of random numbers. The odds of "winning" (and therefore being counted) are set so that over enough trials the quantity is counted on average in accordance to the likelihood of that configuration actually being adopted by the molecules. If many configurations are generated (a million or so), the resulting average correctly reflects the likelihood of each of the possible configurations actually being adopted by the molecules, and a good average for the quantity is produced. Monte Carlo theories of the liquid crystal phase can include various forces between molecules and produce results that sometimes agree extremely well with experimental data.

It should be pointed out that although the fastest computers cannot calculate exactly what happens to a huge number of molecules when they mutually attract and/or repulse one another, we are getting a bit closer. It is now possible for the "exact" motion of each of a "not too large" number of molecules to be followed, taking into account all mutual forces. The configuration of molecules is actually calculated for each of many small steps in time; these configurations are averaged to produce whatever quantity is desired. The chief limitation of this *molecular dynamics* method is that the number of molecules is only around one thousand. Faster computers are sure to drive this number up and produce "exact" predictions starting with molecular forces. Since these calculations start at the atomic level, they have the possibility of determining not only what phases an actual compound possesses but also many of the properties of these phases. In principle, therefore, the properties of an actual compound can be predicted before it is even synthesized, thus saving a huge amount of time by avoiding synthesizing compounds with unremarkable properties. Many calculations of this type are already used for this purpose, but all of them utilize simplifying strategies to reduce the computational complexity.

Can the results of all this theoretical research be summarized? In fact, a few general statements are possible. First of all, indications are that longer range attractive forces and shorter range repelling forces are both important in thermotropic liquid crystals. However, there are reasons to suspect that attractive forces in smectic liquid crystals are more important than in nematic or chiral nematic liquid crystals. The situation is quite different in lyotropic liquid crystals and polymer liquid crystals in solution, where steric forces seem to be by far the most important. Thus once more we see that the liquid crystal phase literally results from a delicate balance between opposing forces. The delicateness of this balance is beautifully displayed in certain experiments done under pressure. As mentioned in the beginning of chapter 8, increases in pressure more or less act like decreases in temperature as far as the stability of certain phases is concerned. Usually, increasing the pressure in the isotropic phase causes a phase transition to the nematic phase and then to the smectic phase. However, in some liquid crystals the sequence upon increasing pressure is isotropic, nematic, smectic, and nematic again. This second nematic phase is called the *reentrant nematic* phase and results because the increase in pressure decreases the intermolecular separation, thus increasing the repulsive steric forces. In certain cases the molecular geometry produces an increase in steric forces large enough to throw the balance back in favor of the nematic phase.

LIQUID CRYSTAL PHASE TRANSITIONS

One of the most exciting areas of both theoretical and experimental physics over the last twenty-five years has been the study of phase transitions. Although we have been talking about phase transitions in which physical properties such as density and index of refraction change, there are many other types of phase transitions in which other physical properties change. For example, the magnetic properties of a solid can change drastically at a certain temperature. Below this temperature the solid can be made into a permanent magnet, whereas this is not possible at a higher temperature.

Likewise, the magnetic and electrical properties of a material change at the phase transition between a normal conductor and a superconductor. What the recent research has told us is that many of these phase transitions have much in common. In fact, some of them are virtually identical, even though the physical properties involved and the underlying mechanisms causing the phase transition are completely different. These results seem to imply that the fact that all these transitions involve large numbers of interacting elements makes many details of the system unimportant. Instead, only a few characteristics of a phase transition determine its nature. There are many different possible liquid crystal phase transitions, so for this reason they have played a significant role in the study of phase transitions in general.

One of the characteristics of a phase transition that is important is the *symmetry* of the phases on each side of the phase transition. Symmetry is an extremely important concept in science and is worth discussing. By symmetry, scientists are referring to what operations can be performed on an object without its changing in any way. For example, consider the objects in figure 10.4. The rectangle shown in figure 10.4(a) can be rotated through 180° about any of the three axes and it will appear exactly as before. Reflection is another operation. Reflection through a plane means all points on one side of the plane are moved perpendicularly through the plane to a point an equal distance away on the other side of the plane. Reflection of the rectangle through the two planes shown also leaves the rectangle unaltered. Note that a square has four axes where 180° rotations leave it the same, but also has one axis where a 90° rotation brings the square back on itself. The square also has four reflection planes instead of two, which have been omitted from figure 10.4(b). Obviously the square has more symmetry than a rectangle.

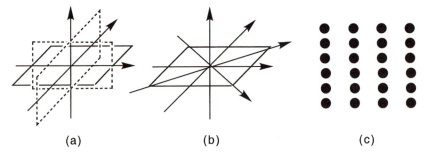

(a)   (b)   (c)

Fig. 10.4. Symmetry operations of a rectangle and square. Three 180° rotation symmetry axes and two reflection symmetry planes (dashed lines) are shown for the rectangle in (a). Four 180° and one 90° rotation symmetry axes are shown for the square in (b). The four reflection symmetry planes are not shown. A rectanglar lattice is shown in (c). Two translation symmetry operations are present in the lâttice in addition to the symmetry operations shown in (a).

Figure 10.4(c) shows a piece of a rectangular lattice. All the symmetry of the rectangle is present in the rectangular lattice, because the lattice is made up of rectangles. But notice that the entire lattice can be translated in certain directions by the right amount and remain unchanged. These translations are therefore also *symmetry operations* for this lattice.

The symmetry of a phase is determined by how that phase can be translated, rotated, reflected, and so on, and still look exactly the same. For example, the isotropic phase has continuous rotational symmetry about every possible axis, in that you can rotate a piece of isotropic liquid about any axis and the phase looks exactly as it did before the rotation. The nematic phase is quite different. There is one axis, the director, about which you can rotate the phase through any angle and not change the phase at all. However, rotation about any other axis usually results in a nematic sample with a director pointing in a different direction. Because the phase is no longer identical after the rotation, we say that continuous rotational symmetry does not exist about that axis. When an isotropic liquid changes to a nematic liquid crystal, some of the symmetry present in the higher temperature phase is lost in the lower temperature phase. This is a general feature of most phase transitions and is referred to as *spontaneously broken symmetry*.

The nematic to smectic transition is another good example of the importance of symmetry. In the nematic phase, there is complete translational symmetry, in that moving the sample through space (without rotation) in any direction does not change the properties at any point fixed in space. This is not true in the smectic phase. Translation parallel to the layer planes keeps the properties at all points the same, but translations perpendicular to the layers can cause properties at certain points in space to change. A symmetry present in the nematic phase is therefore lost in the transition to the smectic phase.

One other important characteristic of a phase transition is the nature of the order parameter involved. Although we have been using the term order parameter to describe the orientational order in liquid crystals, it is really a much more general term. An order parameter can be described for every phase transition. In the higher temperature phase, the order parameter is zero; in the lower temperature phase, the order parameter is nonzero. For example, the order parameter appropriate for the magnetic phase transition is the magnetization (magnetic dipole per unit volume). Above a certain temperature the material has no net magnetization, so the order parameter is zero. At the phase transition temperature the material becomes magnetic, with a nonzero magnetization (nonzero order parameter). In a similar vein, the order parameter appropriate for the nematic–smectic A phase transition must be different from the nematic order parameter S, because S is nonzero both above and below the nematic–smectic A phase transition. The appropriate order parameter must describe the fact that the molecular centers are not uniformly distributed in the smectic phase but instead are bunched into layers. An order parameter that describes a density wave (high density of molecular centers alternating with low density of molecular centers) does the trick. In some transitions (called *discontinuous* or *first-order* phase transitions), the order parameter changes from zero to a finite value at the transition. In other transitions (called *continuous* or *second-order* phase transitions), the order parameter increases from zero gradually at the transition. Some liquid crystal transitions are discontinuous (isotropic-nematic), while other liquid crystal transitions are sometimes continuous (smectic A–smectic C).

This discussion reveals why liquid crystals are important for the study of phase transitions in general. The different liquid crystal transitions afford many possible types of transitions, all with different order parameters. Modern research has revealed that the symmetry of the phases and the mathematical nature of the order parameter are just about all that determine what happens at the phase transition, regardless of what system is under consideration or what properties change at the transition.

DEFECTS IN LIQUID CRYSTALS

A sample of a nematic liquid crystal normally does not possess a director that points in the same direction at all points in the sample. In a certain region of the sample it might point in one direction, and at another point in the sample it might point in a different direction. Often there is a place between these two regions where the director abruptly changes the direction in which it points. It is impossible to define the direction the director is pointing at the point of abrupt change, so this point represents a *defect* in the order of the liquid crystal. Point defects and line defects are the most important types of defects in liquid crystals. As we shall see, the defects present in a liquid crystal sample viewed between crossed polarizers in a microscope determine more than any other factor the appearance of the liquid crystal.

It is important to realize that an abrupt change in the direction of preferred orientation always implies severe distortion of the director configuration in the vicinity of the defect. Previously I discussed how forces are necessary to distort the director configuration, with greater distortions requiring greater forces. These forces would be released if the defect were removed and replaced with a constant director in the region. The fact that defects are stable indicates that outside influences on the director configuration (electric field, nearby glass surface, etc.) are not compatible with a constant director in the region, so a director configuration with a defect is the best compromise possible.

Point defects are less common than line defects. They sometimes do occur in spherical droplets of liquid crystals, around droplets of isotropic liquid in a liquid crystal, and in thin capillary tubes. The most simple example is a spherical droplet of nematic liquid crystal in which the molecules are constrained to be perpendicular to the surface of the droplet. In the center a point defect might occur. As we saw in chapter 6, if the molecules lie parallel to the surface of the droplet, two point defects occur (see figure 6.15). In discussing nematic emulsions in chapter 8, we also learned how a point defect spontaneously develops near a droplet of isotropic liquid in a nematic liquid crystal (see figure 8.12). An interesting example occurs in a thin capillary tube where the molecules must orient themselves perpendicular to the cylindrical surface of the tube. A diagram of a capillary tube is shown in figure 10.5(a), where two point defects are depicted. What is interesting about the two point defects is that the director configurations above the top one and below the bottom one are the same. That means that it must be possible for the two point defects to combine and annihilate each other, resulting in the director configuration (with no defects) shown in figure 10.5(b). The attraction and subsequent annihilation of these two defects is analogous to what happens when two opposite electric point charges attract and "cancel" each other. For this reason, scientists sometimes assign one of these defects a plus sign and the other a minus sign.

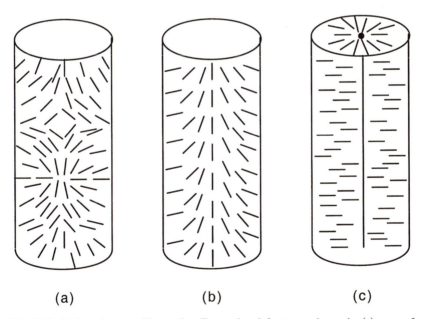

(a)   (b)   (c)

Fig. 10.5. Defects in a capillary tube. Two point defects are shown in (a), one of which is opposite to the other; the director configuration in (b) has no defects; and a line defect is shown in (c).

One other possibility for the capillary tube with the director perpendicular to the surface is that a line defect may be present running along the center of the tube. This is shown in figure 10.5(c). Line defects were given a special name by the early liquid crystal researchers—*disclinations*. The name stems from the fact that the line represents a "discontinuity" in the "inclination" of the director. Many different disclinations are possible. To classify them, we simply examine the director configuration in the plane perpendicular to the direction of the line. Figure 10.6 contains six examples. Notice that signs have been assigned to indicate which disclinations are opposite. The numbers represent the *strength* of the disclination. By comparing figure 10.5(c) to the examples in figure 10.6, you should be able to conclude that the line defect in the capillary tube is in fact a disclination of strength +1.

One important feature of disclinations is obvious from figures 10.5(b) and 10.5(c). A disclination of strength +1 can always relax to the structure of figure 10.5(b) where no defect is present at all. In this sense, a +1 disclination

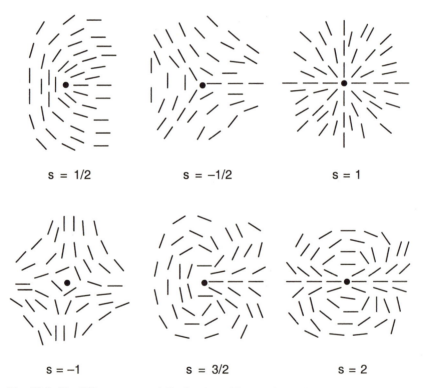

Fig. 10.6. Six different types of disclinations. The numbers give the strength of the disclination. Disclinations of opposite sign can combine to give a disclination of lower strength.

is not absolutely stable. It turns out that this is true for all disclinations of integer strength, whether positive or negative.

As we discussed in chapter 4, places where the director is parallel to the crossed polarizer or analyzer axes appear dark under the microscope. Since defects represent regions where the director configuration changes, these dark regions must also have a unique appearance around defects. To imagine what these dark regions would look like for each of the disclinations shown in figure 10.6, locate the regions around each disclination where the director points either horizontally or vertically. This has been done in figure 4.11 for a disclination of strength +1. Notice that each of the disclinations in figure 10.6 possesses a pattern of dark regions that is unique, allowing for easy identification. A number of the plates contain such characteristic dark areas. Can you recognize them and identify the strength of the disclination?

Defects in chiral nematic and smectic liquid crystals differ from those in nematics due to either the twisted or the layered structure. Chiral nematics and smectics can be deformed in many ways, but the most difficult deformation is one that changes the pitch of a chiral nematic or the layer spacing of a smectic. It is therefore not surprising to find that the defects that do occur in these two types of liquid crystals are the ones that do not involve either of these deformations.

Figure 10.7 illustrates some of the types of disclinations that occur in chiral nematic liquid crystals. The disclination shown in figure 10.7(a) is an important one, since an additional half turn of the chiral nematic helix is introduced at the point of the disclination. This type of disclination shows up in wedge-shaped samples of chiral nematic liquid crystals with the helical axis perpendicular to the surfaces. As the space between the two glass sur-

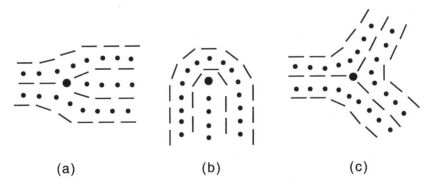

(a)   (b)   (c)

Fig. 10.7. Three types of disclinations in chiral liquid crystals. The dots represent regions where the director points into or out of the page. The lines are regions where the director lies in the plane of the page. The disclination is shown as a large dot.

faces increases, more and more turns of the helix fit in the space between the glass. To keep fitting in more turns of the helix, disclinations must appear at regular intervals in the sample. As discussed in chapter 5, these disclinations are quite visible under the microscope and are called Cano lines. The circular lines in plate 8 are therefore disclinations of the type shown in figure 10.7(a).

Under certain conditions, chiral nematic and smectic liquid crystals form a very interesting texture when placed between two pieces of glass. In both cases, disclinations are responsible for the appearance of this texture. Let us consider smectic liquid crystals in the following discussion, but keep in mind the same description holds for chiral nematic liquid crystals. The only difference is that the layer spacing in the case of smectics is analogous to the pitch in chiral nematics.

Imagine regions of smectic liquid crystal growing in a nematic because the temperature is decreasing. The smectic may be distorted as it grows, but in ways that keep the distance between the layers constant. A typical director configuration is shown in figure 10.8(a). Notice that there are two disclinations running perpendicular to the page, with a disclination in the plane of the page between them. Often the disclinations running perpendicular to the page are actually parts of the same disclination that bends around in an ellipse. This is shown in figure 10.8(b), where another view of the texture is presented. The disclination running through the ellipse (dot in figure 10.8[b]) has the shape of a hyperbola in figure 10.8(a). The presence of these conic sections (plane slices through a cone) is the reason this texture is called the *focal conic texture*.

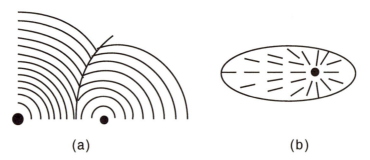

(a)   (b)

Fig. 10.8. Origin of the focal conic texture in a smectic liquid crystal. Diagram (a) shows smectic layers growing from two points on a glass surface. These points are actually parts of a disclination in the shape of an ellipse. Diagram (b) shows the view from above the glass surface, where this elliptical disclination is drawn as a line. The disclination shown as a line in (a) appears as a dot in (b). Plate 4 is a picture of a texture composed of these structures.

FLUID LATTICES

When Reinitzer described the melting behavior of cholesteryl benzoate in 1888, he stated that the substance briefly turned blue as it changed from clear to cloudy. Although researchers working with chiral nematic liquid crystals must have noticed this phenomenon, extremely little appeared in scientific literature for over eighty years. A few reports of experiments were published during the late 1960s and early 1970s, but the situation changed dramatically when detailed experiments revealed that the blue color was due to at least two new liquid crystal phases. Suddenly these *blue phases* became the source of widespread interest, and results soon showed that the structures of these phases were very different from all other liquid crystal phases. New theoretical ideas were born, and entire classes of new experiments were performed on liquid crystals. As we shall see, defects do not just occur at various points in these blue phases; they are essential to its structure.

Before scientists made progress in understanding the blue phases, they had to realize that an old assumption was not always valid. I have already discussed that the helical structure of chiral nematic liquid crystals (figure 1.6) results from the fact that the molecules prefer to be at a slight angle to one another, rather than parallel. For almost one hundred years, scientists assumed that this structure (which involves a single helical axis about which the director rotates) is the most stable structure for such molecules. It turns out that another structure can be more stable. Instead of a single helical axis, the director in this alternative structure rotates in helical fashion about every axis perpendicular to a line. Such a structure has been given the name *double twist*, although an infinite number of helical axes are really present. Figure 10.9 shows the double twist structure. The plane containing the helical axes is the plane of the page in figure 10.9(a). The director points into or out of the page in the center, and rotates as you move outward from the center in any direction. There is no change in the director going into or out of the page. A perspective drawing is shown in figure 10.9(b). The curved lines on the outside of the object represent the fact that the director has rotated 45° in going from the center to the outside.

There is a good reason why this structure does not occur often even though it is more stable than the normal helical (or *single twist*) structure. The double twist structure is more stable only within a small distance from the line at the center. This means that a large double twist structure is not more stable than a large single twist structure. Since this small distance is on the order of the pitch of the chiral nematic (roughly 0.0005 millimeters) and since most samples of liquid crystal are much larger than this, it is not surprising that the double twist structure rarely occurs.

The blue phases are special cases where the double twist structure does fill up a large volume. If you end the double twist structure in all directions from

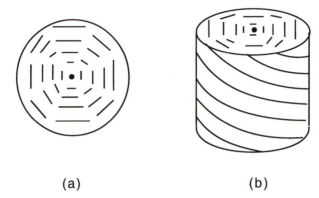

(a)   (b)

Fig. 10.9. Two views of a double twist cylinder. A cross-sectional view (a) shows that in the center the director is parallel to the cylinder axis. The director twists in going away from the center in any direction. As shown in (b), the director has twisted by 45° between the center and outside of the cylinder.

the central line at points where the director has twisted by 45°, a *double twist cylinder* results. This is the structure that appears in figure 10.9(b). Because the radius of this cylinder is small, this cylinder is more stable than an equal volume of single twist chiral nematic. A large structure can be made from these double twist cylinders, but the problem is that disclinations must thread their way between the cylinders. These defects tend to make the structure less stable; yet in one of nature's most delicate compromises, the structure composed of double twist cylinders and defects is slightly more stable than the single twist structure without defects, but only for a temperature interval of roughly 1°C just below the transition from the chiral nematic to the isotropic liquid. Two examples of structures using double twist cylinders are illustrated in figure 10.10.

In the structures of figure 10.10, the disclinations have a cubic arrangement, therefore forming a lattice just as the molecules in a crystal do. There is a big difference between a crystal and this lattice of disclinations, however. In a crystal, a physical entity (a molecule) occupies each lattice point. In these structures, no physical entity, merely a disclination line representing an anomaly in the director configuration, comprises the lattice. The blue phases are these fluid lattices of disclinations.

The reason why crystals reflect X-rays can also explain why a blue phase is capable of reflecting a color (only some blue phases are blue): the side of the cubic cell representing the blue phase disclination lattice is roughly equal to the wavelength of visible light. For reflections of light off the various planes defined by the cubic cell, constructive interference occurs just as it does in crystals where the spacings of atoms are equal to the wavelength of X-rays (see chapter 5). A sample of one of the blue phases will usually have

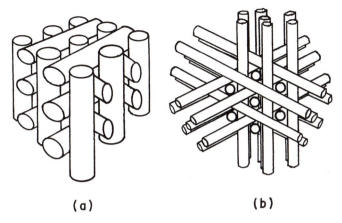

(a)   **(b)**

Fig. 10.10. Two structures composed of double twist cylinders. The orientation of the director at the point where two cylinders touch is the same for both cylinders. In each double twist cylinder, the diractor rotates through an angle of 45° for the structure shown in (a) and 54.7° for the structure shown in (b). Disclinations run through the regions where three cylinders meet.

regions in which the disclination lattice is oriented in different directions. When you observe these blue phases under a microscope using reflected light, these different regions reflect different colors because the wavelength for constructive interference depends on both the spacing between the reflecting planes and the angle between the light beam and the reflecting planes.

These fluid lattices have properties that are extremely similar to crystals. If a substance with a blue phase is cooled extremely slowly so that the blue phase lattice forms very slowly, the lattice will display its cubic arrangement by forming what looks like a single crystal. If the lattice planes are perpendicular to the viewing direction, the single crystal has the shape of a square. Plate 16 is a picture of single crystals of what is called the first blue phase. The cubic lattice of these crystals is turned through an angle relative to the viewing direction, so the single crystals do not have a square shape. In some cases even the facets of these single crystals are visible. It is difficult to look at these pictures and remember that the molecules are not locked into place. Molecules diffuse in and out of the single crystal, but the orientational order of a molecule changes abruptly when it passes from one phase (outside the single crystal) to the other (inside the liquid crystal).

As if two blue phases with two different types of fluid lattices are not enough, nature seems to have come up with a third possible way to find a delicate balance between the double twist structure and defects. A third blue phase exists, yet it does not seem to have the same type of cubic lattice as the other two blue phases. One way to describe this phase is to state that the cubic lattice of disclinations has melted to form a random and fluctuating network of disclinations. At higher temperature the liquid crystalline order necessary to form these disclinations is lost, and the isotropic phase results. A good analogy for this last transition is the liquid to gas transition. In fact, under the proper conditions the third blue phase can change to the isotropic phase without going through a transition, just as is true for a liquid changing to gas above what is known as its critical point.

TWIST GRAIN BOUNDARY PHASES

There is a smectic phase in which the tendency of the molecules to twist spontaneously introduces defects into the structure much as is the case with the blue phases. This phase consists of blocks of smectic A liquid crystal which are separated from each other by *grain boundaries* in which the smectic layers (and the director) discontinuously twist by a small angle. These grain boundaries are separated by a fixed distance along an axis perpendicular to the director, so the effect is to cause the director to twist into a helix that resembles the structure of the chiral nematic phase. This is shown in figure 10.11. Since the grain boundaries are composed of parallel lines of what are called *screw dislocations*, this phase is known as a twist grain boundary phase. Whereas disclinations represent points where the orientational order changes discontinuously, dislocations represent points where the positional order changes discontinuously. In the case of twist grain bound-

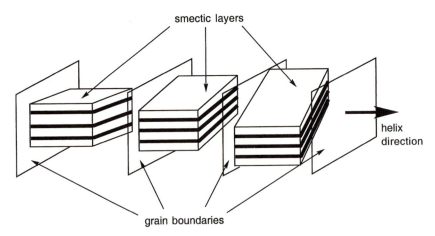

Fig. 10.11. Structure of the twist grain boundary phase that occurs in some chiral smectic liquid crystals. The grain boundaries separate regions in which the smectic layers are at a slight angle relative to one another. The helical axis lies in the plane of the smectic layers and perpendicular to the grain boundaries.

ary phases, it is the positional order due to the smectic layers that changes abruptly at a screw dislocation.

A typical distance between grain boundaries is about 0.025 micrometers, and the amount of twist at each boundary is about 20°. This means that it takes about 360°/20° or 18 grain boundaries for the director to rotate one full revolution. Thus the pitch of a twist grain boundary phase is roughly 18 times 0.025 micrometers or 0.45 micrometers. This places the wavelength of selective reflection, the pitch times the index of refraction, at the long wavelength end of the visible spectrum, so these phases can appear quite colored.

There are other twist grain boundary phases based on the smectic C structure rather than the smectic A structure. Some of the most chiral twist grain boundary phases also form blue phases. That is, for a very narrow temperature range just before the transition to the isotropic phase, these "blocks" of smectic liquid crystal between grain boundaries form double twist cylinders that pack into a cubic lattice. Trying to imagine how the screw dislocations wind their way throughout the structure in a way that produces a cubic lattice of line disclinations is a task that would drive even the most imaginative people crazy!

Chapter 11

BIOLOGICAL IMPORTANCE OF LIQUID CRYSTALS

MOST OF US have heard the statement that biological organisms are 90% water. After hearing such a statement, how many of us then ask how organisms can produce structures that have some degree of rigidity? After all, water has no rigidity at all, yet fairly rigid structures are the norm in the biological world, not the exception. It is important to realize that I am using the word rigid in a very general sense. In addition to the normal meaning of "stiff or unyielding," the word rigid when applied to a biological structure can also describe its ability to confine the movement of material or provide a substrate for the ordering of molecules. The ability of organisms to create rigid structures in an environment rich in water is part of both the development of single cells and the specialization of groups of cells. A single cell, a group of cells performing a specialized function, or an organ are all examples of biological structures, and understanding how such structures form is a central concern of biology. We shall see in this chapter that liquid crystals play a crucial role, not only in the formation of specialized structures but also in the workings of every cell.

BIOLOGICAL STRUCTURES

All biological structures must form in an *aqueous* environment, that is, an environment rich in water. This restriction is actually not that severe. For example, many substances can form crystalline structures in water if the concentration of the substance reaches the saturation point. Chemical reactions in water can produce a product insoluble in water, which then forms a precipitate. Under certain conditions the precipitate forms a rigid structure. Such processes are in fact biologically important, especially in the formation of solid structures, such as bone and shell. Yet these types of structures are not typical of the majority of biologically important structures for a very important reason. Solid structures do not allow molecules to diffuse about so that all the chemical and physical processes necessary for life can take place.

Thus the second requirement of most biological structures is that they be fluid. Only in a fluid do the molecules move from place to place in a way that allows for interactions with other molecules. Think of all the enzymes and proteins in your body. These molecules are produced in one structure, somehow move out of that structure, and are transported to another part of your body, diffuse into another structure, and finally find the right place inside or near a certain cell. For this to occur fast enough, the environment must be highly fluid. This same process takes place on a smaller scale inside each cell. Substances go in and out of the various structures inside the cell. These structures (nucleus, mitochondria, endoplasmic reticulum, Golgi apparatus, etc.) work together to make the cell function properly just as the different organs of the body work together. Thus the ability of biological organisms to form fluid structures of some rigidity in an aqueous environment is central to the existence of life on this planet.

From what we have already discussed, you have probably already guessed that liquid crystals are a natural for such structures. Since the environment is rich in water, all the structures we discussed in chapter 8 should come to mind. Amphiphilic compounds spontaneously form highly ordered structures in a polar solvent such as water, and these structures are liquid crystalline (fluid) rather than solid. This means that there is constant diffusion both of the amphiphilic molecules throughout the structure and of water into and out of the structure. The rigidity of these structures varies over a wide range. Double layers of phospholipid molecules (figure 8.3[b]) are not very rigid at all, while twisted filaments of cylindrical micelles can be very rigid. Both serve extremely important biological needs.

In short, biological structures must be rigid enough to function properly and fluid enough to allow all the necessary processes to take place. There are not many candidates for such structures. We shall see in this chapter that this delicate balance between rigidity and fluidity is achieved in part by liquid crystalline structures.

THE CELL MEMBRANE

The membrane on the outside of every animal cell serves two functions. It confines the contents of the cell to a certain limited volume, and it controls the flow of ions and molecules into and out of the cell. Plant cells also possess a cell membrane, but an additional structure called the cell wall surrounds the cell membrane and provides rigidity in the normal sense of the word. In order for the cell membrane to confine the contents of the cell and maintain the shape of the cell, it must possess some rigidity. To control the transport of substances in and out of the cell, the cell membrane must also be somewhat ordered. A disordered structure cannot discriminate between various substances and therefore is incapable of controlling the movement of certain molecules. Finally, the cell membrane must control the movement of materials and the reactions taking place near the membrane quite rapidly. This can be achieved only if the molecules in the cell membrane can diffuse quickly, which is possible only if the membrane is fluid. Liquid crystalline structures have these characteristics, and the fact that the cell membrane is a liquid crystalline structure is the most important implication of liquid crystals for the field of biology.

The standard model for the cell membrane is called the *fluid mosaic model*. As shown in figure 11.1, the primary structural component of the membrane is a bilayer of phospholipid molecules. Embedded more or less at random in the bilayer are proteins (long sequences of amino acids), which are responsible for the many reactions that occur in the vicinity of the cell membrane. For example, some proteins act as enzymes, while other proteins function as pumps, transporting material into and out of the cell. Some of these proteins lie on the inner or outer surface, and some extend into and through the bilayer. These latter proteins must possess both hydrophobic and hydrophilic sections in order to span the bilayer. The two other main substances found in the cell membrane are cholesterol and water. The cholesterol is found mostly in the hydrophobic region between the polar heads groups, whereas about all of the water is found bound to the head groups. Plant cell membranes contain sitosterol and stigmasterol instead of cholesterol (different steroids) and glycolipids instead of phospholipids (different head groups).

The phospholipids are in the liquid crystal phase, meaning the head groups are not arranged in an array and the hydrocarbon chains are not rigid. The phospholipid molecules diffuse about the cell membrane, as do the proteins (albeit much more slowly). All this motion allows the reactions and interactions necessary for the proper functioning of the cell to take place. At the same time, the ordered structure provides a substrate in which the proteins can order in a specific way, which is also necessary for them to function.

phospholipid

protein

Fig. 11.1. Schematic diagram of a cell membrane, showing the phospholipid bilayer and embedded proteins.

Only certain types of phospholipids are found in animal cell membranes. *Diphosphatidylcholine* is the main ingredient of animal membranes, while *diphosphatidylethanolamine* predominates in the cell membranes of bacteria. As evident from figure 11.2, these two phospholipids have different head groups. In addition, the hydrocarbon chains usually have an even number of carbon atoms, with 16-, 18-, and 20-carbon chains accounting for 98% of the phospholipids in animal membranes. The majority of the bonds between the carbon atoms are single bonds, where the carbon atoms share one pair of electrons. Double bonds, in which two electron pairs are shared, are sometimes present.

Ingenious experiments have demonstrated just how crucial the liquid crystalline nature of the bilayer is to the proper functioning of cells. In chapter 8, I discussed that phospholipid bilayers possess a phase called the gel phase at temperatures below the liquid crystal phase. At the transition from the liquid crystal phase to the gel phase, the fluid hydrocarbon chains straighten, and the polar head groups arrange themselves into a hexagonal array. The experiments consist of simply allowing a primitive organism to grow at a certain temperature and after a while extracting the membranes to examine at what temperature the liquid crystal to gel phase transition occurs. If the organism

(a)

(b)

Fig. 11.2. Two typical phospholipids found in cell membranes: (a) diphosphatidylcholine, and (b) diphosphatidylethanolamine.

is forced to grow at a temperature of 35°C, the membranes have a transition temperature slightly below 35°C. If the organism grows in a 30°C environment, the membranes possess a transition temperature slightly below 30°C. Since the membranes have a transition temperature just below the ambient temperature, the membranes are all in the liquid crystal phase. Lowering the temperature below the transition temperature always changes the functioning of the cell membrane and often leads to the death of the organism. Work done with bacteria that live in the deep ocean trenches, where the pressure is much greater than atmospheric pressure, illustrates the need for liquid crystalline cell membranes. High pressure tends to raise the transition temperature. Organisms growing under high pressure therefore produce cell membranes that have transition temperatures just below the ambient temperature when under high pressure, but have transition temperatures far below the ambient temperature when the pressure is reduced to atmospheric values. Many of these organisms therefore die when brought to the surface of the ocean. This implies that it is also important that the gel to liquid crystal transition temperature not only be below the ambient temperature but also be fairly close to the ambient temperature.

How can an organism control the transition temperature of its cell membrane? You may recall that different liquid crystals possessed different phase transition temperatures. Phospholipids with the same polar head group have different transition temperatures depending upon the length of the hydrocarbon chains and the number of double bonds in the hydrocarbon chains. The shorter the hydrocarbon chains and the more double bonds present in the chains, the lower the transition temperature. Biological organisms select the proper phospholipid molecules during growth so that the transition temperature of the bilayer occurs just below the ambient temperature.

Although the physical changes due to the presence of cholesterol in cell membranes are known, the biological importance of these effects is still not understood. In general, the addition of cholesterol to a phospholipid bilayer lowers the gel to liquid crystal phase transition temperature, reduces the sharpness of this transition, and changes the rate at which molecules diffuse about the bilayer. As the cholesterol concentration rises, there also seems to be a tendency for the proteins of the cell membrane to cluster rather than to distribute themselves randomly.

It should be pointed out that cell membranes are not symmetric. That is, the outside is different from the inside. For example, some proteins lie more or less on one of the surfaces of the membrane. The proteins present on the outer surface are different from those on the inner surface. Even when the same protein is present on both surfaces, its orientation (and therefore its function) is not exactly the same on both surfaces.

There is good evidence that the phospholipid bilayer is important for the proper functioning of the proteins in the cell membrane. Proteins perform their various functions by folding up their long sequence of amino acids in a very specific way. An extremely slight rearrangement of how the protein is folded is sometimes all that is necessary to stop the protein from performing a certain function. Interactions between the proteins and the phospholipid molecules affect how the amino acid sequence of a protein is folded and therefore affect how the protein functions. We know these interactions occur, because experiments show that the phospholipid molecules in the vicinity of a protein are more highly ordered and not as free to diffuse as the phospholipid molecules in the rest of the membrane.

OTHER STRUCTURES

The outer cell membrane is only one of the many bilayer membranes which occur in a cell. Membranes within a cell serve to compartmentalize various functional units within the cell. For example, the nucleus of a cell is surrounded by a phospholipid membrane that is very similar to the outer cell membrane. Slight differences exist between the types of phospholipids contained in these internal membranes, and obviously some of the proteins present are different. For example, the energy for a cell to function is delivered to parts of the cell by a substance called *adenosine triphosphate* (ATP), which is synthesized on the internal membrane surrounding the mitochondria. Similarly, photosynthesis takes place in plant cells on the internal membrane surrounding the chloroplasts. One last example of an internal membrane is the *endoplasmic reticulum*. This is a long, flattened bilayer structure to which the *ribosomes*, where RNA is manufactured, are attached.

From this partial list of important cell structures, it becomes clear that the phospholipid bilayer is the chief means by which different structures within a cell are segregated from the rest of the cell. In this sense, this liquid crystalline structure is one of the main building blocks of specialized structures within animal and plant cells.

The functioning of specialized cells is sometimes due to liquid crystalline structures. The visual cells of vertebrates contain stacks of flat bilayer vesicles that contain the molecules which are sensitive to light. The high sensitivity of these cells to light is achieved by the high concentration of these molecules that results from this stacked arrangement. Impulses are carried along highly elongated nerve cells, which are surrounded by a cylindrical sleeve made of *myelin*. This myelin sleeve is made of cells (with protein-containing phospholipid bilayers) that spiral around the nerve many times. Its function is to separate the nerve cells from the surrounding environment, so that they can perform their function. In this sense, the myelin sleeve functions in the same manner as insulation around a wire. As mentioned in chapter 2, the forms these sleeves adopt in water when viewed under a microscope (called *myelinic forms*) were noted as early as 1850.

Many important biological functions are possible due to the ordered arrangement of microfilaments and microtubules in cells. *Microfilaments* are long slender structures consisting of certain proteins that stack together. In this sense these microfilaments are analogous to polymers, except the "monomer" in this case is a protein molecule. The shape and movement of cells is controlled by an oriented array of these microfilaments just under the cell membrane. Muscle cells are packed with two different types of microfilaments, which arrange themselves in a way reminiscent of nematic and smectic liquid crystals. The more slender *actin* filaments lie between the thicker *myosin* filaments as shown in figure 11.3. The myosin filaments contain a massive protein that, if activated, tends to associate more closely with the actin filaments. Activation of this protein on all the myosin filaments causes contraction of the muscle cells, as depicted in figure 11.3(b). In striated muscle (for example, in arm and leg muscles), the actin filaments slide relative to the myosin filaments. Although the contraction is very rapid, it is limited in both extent and duration. In smooth muscle (heart muscle, for example), the ordering of the actin and myosin filaments is more nematic than smectic. This arrangement is shown in figure 11.3(c). During the contraction of these cells, the actin filaments move relative to the myosin filaments, and the myosin filaments move relative to each other.

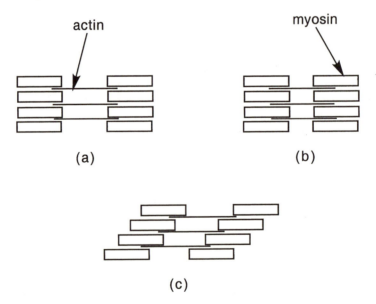

Fig. 11.3. Schematic diagram of muscle fiber, showing the actin and myosin filaments. Striated muscle is shown in its (a) relaxed and (b) contracted state. The arrangement of the filaments in smooth muscle is shown in (c).

This produces a greater amount of contraction, and the contracted state can be maintained for longer periods of time.

Microtubules are similar to microfilaments but are larger, hollow cylinders of linearly stacked proteins instead of smaller, solid cylinders. Microtubules can be quite rigid, due to the spiral ordering of the proteins. Microtubules are thought to form the *cytoskeleton* of cells, that is, the structure responsible for cell size and shape. Microtubules also are important for some of the slow movements that occur within cells. For example, during *mitosis* (cell division), microtubules called spindle fibers form between the chromosomes and one of the two *centrioles*. The chromosomes move along these microtubules and thus segregate, with half of the chromosomes around each centriole. This process is shown in figure 11.4. The formation of a chromosome pair for each of these chromosomes and the introduction of a cell membrane then completes the process of cell division.

The arrangement of DNA in chromosomes seems to have a liquid crystalline character. The long strand of DNA winds itself back and forth in planes perpendicular to the long axis of the chromosome and works itself along this axis by moving from one plane to the next. The direction of the winding of the DNA rotates along the long axis of the chromosome, just as the preferred direction in a chiral nematic liquid crystal rotates throughout the material. Near the side of a chromosome, the DNA strand must turn around and head back into the chromosome. As previously discussed, the double helix structure is fairly rigid, so the 180° turn cannot be extremely tight. The implication of this is that the side of a chromosome cannot be very sharply defined. Loops of the DNA strand extend into the region around the chromosome. The side of a chromosome therefore resembles the interface

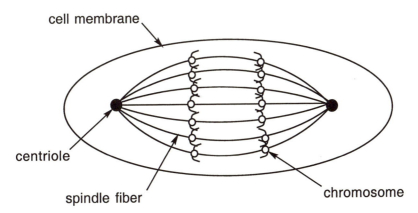

Fig. 11.4. A cell undergoing cell division. The chromosomes move along spindle fibers toward the two centrioles.

between a liquid crystal phase and an isotropic phase. The strand of DNA is highly ordered inside the chromosome and fairly disordered outside the chromosome. As before, however, one must keep in mind that both phases are fluid. The DNA strand is free to move or change its orientation, but when it is in the chromosome it maintains an average orientation in concert with the rest of the DNA strand. When outside the chromosome, it is much less constrained and fluctuates with little orientational or positional order on average.

There is also evidence that the liquid crystalline ordering of polymer material and proteins is the first step in the creation of very hard biological structures such as cuticle and bone. The ordering seems to be in layers, with the preferred direction in each layer rotating slightly in going from layer to layer. At first the order is liquid crystalline, but as the concentration of polymer and protein increases, the material becomes more of a gel. Crystalline bundles of parallel polymers also can form, causing the material to harden. Such a substrate seems to be necessary for minerals to begin to accumulate, thus allowing shells, bones, and teeth to form.

The clotting of blood involves the production of long strands of *fibrin*, which form a close mesh to trap red blood cells. The fibrin forms from the protein *fibrogen* that is present in blood. The complex reactions that take place when a blood clot forms cause the fibrogen to polymerize and form long ordered fibrin strands.

DISEASE

If liquid crystalline structures are essential for many biological processes, conditions that change the liquid crystalline nature of these structures can have deleterious consequences. For example, the disease multiple sclerosis is characterized by the local disintegration of the myelin sheath around nerve axons. Many factors have been implicated as possible causes of this disease, with all of them somehow having an effect on the stability of the spiral cell structure of the myelin sheath. Basic research continues to investigate the physical and chemical conditions under which such a structure is stable, in hopes of tracking down a firm explanation for this condition.

As I previously discussed, liquid crystalline structures control the size and shape of cells. Normally, red blood cells are disc-shaped, with the center region thinner than the outer region. This shape possesses a large surface area, which promotes the rapid exchange of gases between the cell and the surrounding plasma. The blood of people suffering from sickle-cell anemia contains red blood cells that are distorted into a sickle shape. Not only does this shape hurt the efficiency of gas exchange, but these cells cause the viscosity of the blood to increase, thereby reducing blood flow throughout the body. The cause of sickle-cell anemia is hereditary and is due to the production of abnormal hemoglobin. This form of hemoglobin possesses a higher viscosity than normal hemoglobin, due to its tendency to form more ordered phases of the long, thin molecules. The presence of these more ordered structures tends to stretch out the red blood cells, causing them to adopt a different equilibrium shape. Either directly or indirectly, this causes a more sluggish flow of blood. An understanding of the physical characteristics of both forms of hemoglobin is certainly called for as research into sickle-cell anemia continues.

Atherosclerosis is another disease in which liquid crystalline structures are implicated. In this disease, local thickening of the walls of arteries supplying blood to the heart, brain, and other vital organs obstructs the flow of blood. Investigation of the lesions that cause the thickening reveals that they contain a much higher percentage of cholesterol esters than healthy artery walls. The difference between cholesterol and cholesterol esters is important. Cholesteryl myristate is a cholesterol ester (see figure 1.10). Cholesterol is a shorter molecule, with only an HO group instead of the $C_{13}H_{27}COO$ group of cholesteryl myristate. Many cholesterol esters are liquid crystalline at body temperatures, whereas cholesterol has no liquid crystal phase whatsoever. Microscopic studies of diseased artery walls have revealed droplets with the classic "brushes" in polarized light, reminiscent of liquid crystal droplets. It is clear that some of the cholesterol esters that build up in these lesions are in the liquid crystalline phase. The reason for the buildup of cholesterol esters is not understood. There seems to be a steady increase in the amount of cholesterol present in artery walls with age. Only so much cholesterol can be present in combination with water and phospholipids before the cholesterol begins to crystallize out of solution. The creation of cholesterol ester droplets in which cholesterol can dissolve may simply be a way to contain the high amount of cholesterol without it crystallizing out. In advanced cases of atherosclerosis, the lesions actually contain crystalline cholesterol. The problem is that the rate of exchange of various substances between the lesions and the blood is fairly slow, meaning that it is difficult for the body to dissolve cholesterol or its esters once they accumulate. Knowledge of the phases present in these lesions is important to those trying to find ways to reverse the condition. As long as the cholesterol is contained in liquid or liquid crystalline phases, there is a chance that it can be removed through metabolic processes. Crystalline cholesterol presents a far greater problem, as interactions between the cholesterol molecules and surrounding substances take place at a much reduced rate. Some researchers are investigating ways to remove cholesterol from cholesterol ester droplets as a possible method to reverse the progress of the disease.

Liquid crystalline structures also seem to play a role in cancer. The microfilaments next to the cell membrane of malignant cells seem to be less ordered than in normal cells. In a normal cell, the organization of these microfilaments is coordinated around the cell surface. In this way the movements of the cell are controlled, and the cell properly responds to the presence of neighboring cells. When a normal cell becomes malignant, the coordination of the orientation of these microfilaments is lost. In some areas of the cell, the microfilaments are very disordered. These regions form irregular features that move in uncontrolled ways. The result of these uncontrolled movements of regions of the cell surface is that these cells tend to invade surrounding tissue.

The reason for this disorganization of microfilaments is unknown, but the entire cell membrane is involved. For example, one of the first signs that a cell has become malignant is that the adhesiveness between it and other cells is reduced. This shows up as small gaps between the malignant cell and surrounding cells. A reduction of the interactions between cell membranes involves the proteins on the outer surface, indicating that malignant cells have also changed the composition or structure of these outer surface proteins. The membranes of malignant cells usually contain more cholesterol than those of normal cells, so as we have already seen, this changes the amount of order in the bilayer structure. There is some evidence that proteins in the cell membranes of malignant cells tend to cluster more than in the membranes of normal cells. Whether this is simply a result of the change in order within abnormal cell membranes or symptomatic of something more directly involved is an open question.

ANESTHETICS

An interesting example of the importance of liquid crystalline order in cell membranes concerns anesthetics and how they might act to control pain. One class of local anesthetics is thought to inhibit the action of certain proteins that extend from one side of the cell membrane to the other. One piece of evidence for this is that the more soluble the anesthetic is in the cell membrane, the more effective it is in controlling pain. One hypothesis for how this occurs is that the anesthetic molecules disrupt the membrane order in a way that changes the folding of the protein and thus renders it ineffective. A slightly different hypothesis is that the anesthetic molecules are most soluble in the part of the phospholipid bilayer near proteins, where the order is slightly different from the rest of the bilayer. Again, by disrupting the normal order around the protein, the protein's normal function is lost. The nice aspect of this last suggestion is that it explains why some anesthetics work so well even when the dose is quite low.

SUMMARY

What is clear is that the delicate properties of the liquid crystal phase are important to many biological processes. The normal functioning of biological structures depends on the compromise between order and disorder that characterizes all liquid crystals. While this delicate response is essential for controlling all the interactions and responses taking place, it also suffers from being extremely vulnerable to anything that even slightly changes the environment. This delicate balance may be upset by the addition of a new substance, by a slight change in a normal substance, by a small change in the concentration of ions, or even by a small change in temperature, causing a finely tuned system to malfunction. In some areas of biology, researchers are starting to ask questions about the physical state of important molecules, in hopes of uncovering important new insights into how organisms perform necessary functions. It seems quite fitting that in many cases they will be dealing with liquid crystals, nature's delicate phase of matter.

SUGGESTIONS FOR FURTHER READING

ALMOST ALL the literature devoted to liquid crystals has been written for specialists in the field. For this reason, all the books listed below are at a more advanced level than this book. Although the reader of this text should have acquired a background in liquid crystal chemistry and physics sufficient to understand the more advanced treatments used in these books, the mathematical development and scientific terminology may cause problems.

Bahadur, B., editor. *Liquid Crystals: Applications and Uses*. Volumes 1, 2, and 3. World Scientific Publishing, 1990.

Chandrasekhar, S. *Liquid Crystals*, 2nd edition. Cambridge University Press, 1992.

Collings, P. J., and J. S. Patel, editors. *Handbook of Liquid Crystal Research*. Oxford University Press, 1997.

Collings, P. J., and M. Hird. *Introduction to Liquid Crystals: Chemistry and Physics*. Taylor & Francis Ltd., 1997.

deGennes P. G., and J. Prost. *The Physics of Liquid Crystals*, 2nd edition. Oxford University Press, 1993.

Demus, D., J. Goodby, G. W. Gray, H.-W. Spiess, and V. Vill. *Handbook of Liquid Crystals*. Volumes 1, 2A, 2B, 3, and 4. Wiley-VCH Verlag GmbH., 1998.

Gray, G. W., and J. W. G. Goodby. *Smectic Liquid Crystals: Textures and Structures*. Leonard Hill, 1984.

Kumar, S., editor. *Liquid Crystals in the Nineties and Beyond*. World Scientific Publishing, 1995.

Vertogen, G., and W. H. deJeu. *Thermotropic Liquid Crystals, Fundamentals*. Springer-Verlag, 1988.

White, A. M., and A. H. Windle. *Liquid Crystalline Polymers*. Cambridge University Press, 1992.